Resonant Matter

New Approaches to Sound, Music, and Media

Series Editors: Carol Vernallis, Holly Rogers, and Lisa Perrott

Forthcoming Titles:

Biophilia by Nicola Dibben
David Bowie in Music Video by Lisa Perrott
Animated Music Notation by Cat Hope and Ryan Ross Smith
Popular Music, Race, and Media since 9/11 by Nabeel Zuberi
Popular Music and Narrativity by Alex Jeffery

Published Titles:

Transmedia Directors by Carol Vernallis, Holly Rogers, and Lisa Perrott
Dangerous Mediations by Áine Mangaoang
Resonant Matter by Lutz Koepnick

Resonant Matter

Sound, Art, and the Promise of Hospitality

Lutz Koepnick

BLOOMSBURY ACADEMIC
NEW YORK • LONDON • OXFORD • NEW DELHI • SYDNEY

BLOOMSBURY ACADEMIC
Bloomsbury Publishing Inc
1385 Broadway, New York, NY 10018, USA
50 Bedford Square, London, WC1B 3DP, UK

BLOOMSBURY, BLOOMSBURY ACADEMIC and the Diana logo are trademarks of Bloomsbury Publishing Plc

First published in the United States of America 2021

Cover design: Louise Dugdale
Cover image: Ragnar Kjartansson: The Visitors, 2012. Nine-channel video. Duration: 64 minutes. Commissioned by the Migros Museum für Gegenwartskunst, Zurich. Photograph: Elísabet Davids. © Ragnar Kjartansson; Courtesy of the artist, Luhring Augustine, New York and i8 Gallery, Reykjavik

Library of Congress Cataloging-in-Publication Data
Names: Koepnick, Lutz P. (Lutz Peter) author.
Title: Resonant matter: sound, art, and the promise of hospitality / by Lutz Koepnick.
Description: New York City: Bloomsbury Academic, 2021. |
Includes bibliographical references and index. | Summary: "The first overview of resonance/vibration - covering the field thematically, and written by one of the leading figures in sound studies and sound art"–Provided by publisher.
Identifiers: LCCN 2020039605 | ISBN 9781501343377 (hardback) | ISBN 9781501343384 (epub) | ISBN 9781501343391 (pdf)
Subjects: LCSH: Sound art–History and criticism. | Sound art–Philosophy and aesthetics. | Sound (Philosophy) | Voice (Philosophy) | Resonance–Philosophy. | Video installations (Art) | Ragnar Kjartansson. Visitors.
Classification: LCC ML3295 .K64 2021 | DDC 781.1/7–dc23
LC record available at https://lccn.loc.gov/2020039605

ISBN: HB: 978-1-5013-4337-7
PB: 978-1-5013-4367-4
ePDF: 978-1-5013-4339-1
eBook: 978-1-5013-4338-4

Series: New Approaches to Sound, Music, and Media

Typeset by Deanta Global Publishing Services, Chennai, India

Contents

Illustrations

Figures

Plates

Prelude

The questions that set this book in motion emerged with startling urgency in late 2016 during my fourth or fifth viewing of Ragnar Kjartansson's *The Visitors* at the Frist Art Museum in Nashville: Why does the Icelandic artist's nine-channel sound and video installation resonate with so many of its audiences and, as an NPR feature in October 2016 had it, move them to tears, over and over again?[1] And how is it possible that *The Visitors*' at once deeply melancholic and playful sounds manage to cause such strong attachments in spite of the fact that not much seems to happen during its sixty-four-minute loop?

The initial plan was to write a very short, very personal, and what I believed to be very straightforward book. It was to offer some thoughts about how music touches listeners, spookily yet effectively, at a distance. Its central ambition was to ask whether and how sound can push against the walls of ceaseless self-management we call neoliberalism. What you have in front of you still does all this, but the path necessary to answer my two questions turned out to be much longer, winding, and challenging than originally anticipated. It took me to places, ideas, concepts, and other artworks that were not part of the original vision. And it resulted in a book that, in order to understand how and why *The Visitors* resonates with listeners over and over again, cannot help but to explore the nature of resonance itself, of hospitality amid inhospitable times; to think about art's unique invitation to coexist with things strange and other; to ponder about the role of emotional attachments in face of the devastating landscapes of the Anthropocene; to address the vibrancy of matter and the pleasures of sensing sounds that exceed our hearing; and to discuss the borders we erect to defend our lives and the ones we need to tear down to secure possible futures. Each chapter of what is to follow pivots around a different feature of *The Visitors* and its ecology of sounds and images. Yet all approach Kjartansson's installation as a hospitable site to pursue a wide range of inquiries, bring into play various other works of contemporary sound and installation art, and precisely thereby develop what it takes to answer the initial questions.

[1] https://www.npr.org/2016/10/28/498718095/art-star-ragnar-kjartansson-moves-people-to-tears-over-and-over.

The concepts of hospitality and resonance feature as two central terms of analysis throughout this book. The first demands categorical openness toward what is foreign, different, and unexpected while the second describes causal effects and affective relationships that often exceed a distant observer's impartial grasp. The following pages make a case that both concepts belong to each other. One needs and energizes the other. Each illuminates the other's shape. And both matter as perhaps never before amid a world whose ever-more inhospitable landscapes of climate change, neo-tribal walling, and restless self-optimization undo the ground for resonant interactions, for acts of mimetic playfulness, aesthetic surrender, and empathetic entanglement, for attachments that transcend the rhetoric of control and strategic self-assertion. Hospitality and resonance ask us to attune to and coexist with what is not us. They pierce the confines of our imagination, unsettle intended outcomes and predictable results, and cause us to reckon with what defies instant recognition. Generous hosts welcome visitors whether or not their arrival suits the host's timetables, convenience, or spatial affordances. Resonant objects echo the vibrations of other matter, yet such echoes often render a nearby source's sounds strange and animate what at first may appear mute and lifeless. The demands of both hospitality and resonance foil the surfeits of intentionality, goal-oriented action, and interpretation. They make humans and nonhumans, subjects and objects, follow detours rather than proceed in straight lines. They sway us to pursue unforeseen routes and slow down our strides rather than jump on to the next thing. They ask us to attend to the ripples we set in motion when throwing a rock into a pond rather than focus on the hand that initiated the process.

Method is detour, a German critic once wrote, possibly implying that the opposite might ring true as well.[2] And detours are part of *Resonant Matter*'s critical method indeed. Each of the main chapters of this book explores different aspects of Kjartansson's *The Visitors* and its aesthetic of resonant hospitality. And yet, in order to do so, *Resonant Matter* engages with the work of many other artists as well, some of them well covered by previous critics already, others not so. We, for instance, revisit seminal performances by Alvin Lucier and Karlheinz Stockhausen to place Kjartansson's play with resonant sounds and electric transduction in larger historical contexts. The work of David Rothenberg and Marcos Ávila Forero helps us to focus on the role of water and *The Visitors*' iconic bathtub, while artists such as Tanya Tagaq, Juliana Snapper, and Camille

[2] Walter Benjamin, *Gesammelte Schriften*, vol. 1.1 (Frankfurt am Main: Suhrkamp, 1974), 208.

Norment help elucidate the vocal performances of Kjartansson's band. You'll hear about recent installations by Teresa Margollis and Mischa Kuball as I explore the extent to which sound in contemporary art may reroute the impact of painful pasts; and by Philip Lachenmann and Carsten Nicolai as we follow the paths of resonant vibrations into subatomic realms that exceed the visible. Last but not least, we take a detour through the work of artists as diverse as Susan Philipsz, Lawrence Abu Hamdan, and Guillermo Galindo in order to better understand the political index of *The Visitors*, the method behind its at first perhaps most striking madness: the gathering of eight musicians across discrete rooms, trying to tune into each other's sounds in spite of the walls that separate them.

Resonant Matter's various detours may cause some readers to think of this book either as a close reading of a single artwork unable to sustain focus or as a survey of the domains of contemporary sound and installation art unwilling to offer the full picture. It is both at once and therefore, I hope, neither. It aims at something that I in the process of writing began to call resonant criticism: a mode of deliberation that follows the lead of its objects and their entanglements with other objects and practices at the risk of being seen as non-linear, subjective, eclectic, and partial. While each chapter, like a rude guest overstaying a host's patience, will return to Kjartansson's installation, the book as a whole isn't meant to provide a monograph meticulously mapping *The Visitors*' influences and genealogies, its exact place in the contemporary art world, and its aesthetic, political, economic, or ideological echoes. Moreover, while you will read about many artists experimenting with figures of resonance and vibrational transduction, no attempt is being made to define a canon of contemporary sound art, a term whose legitimation is controversial even among some of the practitioners discussed here. Resonant criticism explores the possible copresence, on the one hand, of different artistic positions, texts, works, and interventions in the space of our thinking about art and, on the other, of diverse modes of hearing, listening, attending, and attaching ourselves to—of sensing—music and vibrating matter. It tunes our writing and thinking to the sounds (and as you will see: the unsound) of aesthetic events such as *The Visitors*, as much as it seeks to detect and establish echoes between different works and ecologies of hearing so as to account for their specificity. To "read" here means to think through the routes along which one thing possibly leads to another and may cause infectious vibrations. It is to invite diverse guests into the house of writing and appraise their interactions, their likely and unlikely exchanges and effects on each other. As so often in academic writing the proof here, too, is in the pudding: how constellations of materials allow us to see and understand aesthetic materials in new and unexpected ways.

A good friend and colleague asked me with genuine curiosity late in the process how I would quote *The Visitors*, a work whose nine screens have been installed quite differently during its many exhibitions since its premiere in Zurich in 2012 and whose spatial choreography defies the idea of a unified spectator able to take everything in all at once. No "reading" of *The Visitors* can ever come in the singular, nor does any statement about how we may want to read it in the first place. The scholarly practice of using citations as evidence for truth claims clearly hits its limits when we approach aesthetic objects as events rather than texts and ask our writing to coevolve—resonate with—the thinking of its objects. Resonant criticism does not argue for or against its objects of study. It argues along and with them. It is neither driven by deep-seated suspicion about the affective force of the aesthetic nor does it seek to persuade the reader that one reading is simply better and more persuasive than another. To interpret, here, does not mean to assess different meanings and—with the support of well-placed quotations—privilege one over the other. Following the original Latin meaning of interpretation, it instead aspires to engage in and be engaged by what I understand to be an open-ended and reciprocal process of translation and transduction, call and response, event and echo.

As I recall in Chapter 1 of the book, the concept of resonance has often been seen as the province either of mechanistic theories of matter or the capriciousness of affect. You may skip this more introductory chapter if you want to cut right through to *The Visitors*, but if you are interested in why later chapters embrace ambivalent understandings of resonance as the concept's true strength, I recommend not jumping ahead after all. For resonance's multiplicity, its curious blend of determinism and play, causality and contingency, provides the pages to come with an effective model of how to think about both the specificity of art and our critical engagement with it. Just as art cannot do without subjecting us to something that is not us or of our own making, yet only in so doing succeeds in energizing new thought, so does resonant criticism approach its objects not from a position of detachment and exteriority but of irreducible entanglement. It "reads" and "quotes" installations such as *The Visitors* similar to how a seismograph charts the tremors of a distant and barely sensible earthquake. It considers meaning not as what readers may find in a given text, but in terms of the ripples, echoes, and reverberations, the interferences and unintended consequences, that emanate from it. In analogy to the gestures of unconditional hospitality, resonant criticism invites events and their (mediated) memories to inhabit the space of our speaking, writing, and thinking about art in order to reflect upon what it may mean to live with aesthetic objects, events, and materials beyond our initial encounters.

Because *Resonant Matter* wants to think along the movements of art and the vagaries of our listening and seeing, much of recent excellent writing on resonance by sociologists, anthropologists, architects, philosophers, legal scholars, and pedagogues has receded somewhat into the background.[3] The decision was a conscious one, yet it should not be seen as a gesture of disrespect for all of this work. On the contrary. I have greatly profited from the work of numerous scholars of resonance, yet I refrain from staging prolonged conversations with this writing on its own terms because it would have sidelined the effort to attend to and probe the thinking of art itself. This is not to say that *Resonant Matter* entirely avoids theoretical deliberation. On the contrary. Chapters such as the ones on the vocal dynamics of *The Visitors*, on the temporalities and durations of Kjartansson's performances, on the posthuman inflections of resonant aesthetics, and on the role of subatomic vibration in quantum mechanics include many more theoretical moves than I initially anticipated. What it is to say, however, is that this book wants to allow "theory" primarily to emerge from its objects and their echoes in our listening rather than to unleash prefabricated concepts from a point of intellectual mastery onto sonic matter. I am certain some readers will fault me for my decision not to address numerous philosophical traditions that could have been mobilized to think through the (a)logic of resonant vibrations and interactions in contemporary sound art, while a good number of other readers will no doubt lack extended passages situating the work of different artists presented here in their respective cultural and art historical contexts and question the inclusion of some at the costs of excluding other artists. So, let me be clear upfront. If you are looking for a book offering either a comprehensive theory or a textbook history of sound art, this one is not really for you. If you instead are curious about how contemporary sound and installation art can energize our thinking about ecologies of hospitality amid the Anthropocene's

[3] Think, for instance, of the work of Hartmut Rosa, *Resonance: A Sociology of Our Relationship to the World*, trans. James Wagner (Cambridge, UK: Polity Press, 2019); John M. Meyer, *Engaging the Everyday: Environmental Social Criticism and the Resonance Dilemma* (Cambridge, MA: MIT Press, 2015); David Farland, *Drawing on the Power of Resonance in Writing* (Scotts Valley: CreateSpace, 2013); Peter Price, *Resonance: Philosophy for Sonic Art* (New York: Atropos Press, 2011); and—somewhat addressed in Chapter 1—Veit Erlmann, *Reason and Resonance: A History of Modern Aurality* (Cambridge, MA: MIT Press, 2010); and of various essays gathered in *Milieux Sonores / Klangliche Mileus: Klang, Raum und Virtualität*, ed. Marcus Maeder (Bielefeld: Transcript, 2010).

inhospitality; about resonant relations of mind and matter, solitariness and communality, the human and the nonhuman; about the politics of voice and the transformative power of listening; about sound's ability to address painful pasts and reroute traumatic histories; and, last but not least, about contemporary art's capacity, as art, to speak up against the neoliberal geopolitics of walling and division, then this book, I hope, will be up your alley.[4]

The writing of this book required many physical and metaphorical voyages to reach its final shape. After listening to *The Visitors* more often, I presume, than anyone else during its showing at the Frist, I had the opportunity to test some of my initial thoughts about Kjartansson's installation in the museum's *Food for Thought* series with chief curator Mark Scala. Thanks go to Mark and his staff for enabling this conversation and urging me to bring some order into my rather diffuse set of initial responses. Later trips to Washington, DC, New York, Annandale-upon-Hudson, Venice, Berlin, Havana, Kuala Lumpur, and Singapore were as important to research this book as had been—rather unknowingly at that time—earlier travels to Iceland and the United Kingdom. I am thankful to Vanderbilt University for supporting my research over the years, and to many of my undergraduate and graduate students patiently hearing more about this project than, I am sure, they had bargained for. Their feedback was invaluable along the passage of writing this book, as were audience responses to lectures and workshops I held about figurations of resonance in contemporary media art in Aarhus, Atlanta, Basel, Chapel Hill, Cologne, Münster, Nashville, Portland, Oakland, and Zurich.

I am deeply grateful to Carol Vernallis, Holly Rogers, and Lisa Perrott, editors of the New Approaches to Sound, Music, and Media series, for taking on this project, urging me to adjust several of its premises, and encouraging me to experiment with what I considered "new." Special thanks go to Leah Babb-Rosenfeld and Rachel Moore at Bloomsbury for patiently ushering the project toward publication; to Caroline Burghardt at Luhring Augustine in New York for her tremendous help with securing many of the images reproduced in this volume; and to Jackie Olson back at Vanderbilt for tracking down various copyrights. Nothing of what you have in front of you, however, would have been

[4] The manuscript for this book was completed before COVID-19 turned social distance and isolation into a prolonged and taxing order of the day. It was tempting during the book's final production to add various passages exploring the role of resonance and hospitality amid a world seeking and pushing against shelter in place. In the end, I decided against it, confident the readers of the book's chapters will draw out their own connections between *The Visitors*, resonant art, and the pandemic's lasting effects on the fabrics of sociability.

possible without the inspiration of my most trusted travel companion in life, Jana Harper. Her infectious sense of generosity and hospitality animates each and every page of this book. The privilege of sharing our itineraries, thoughts, and passions—our lives—is what resonance in the end is all about. It is with resounding love and gratitude that I dedicate *Resonant Matter* to her.

1

On Resonance

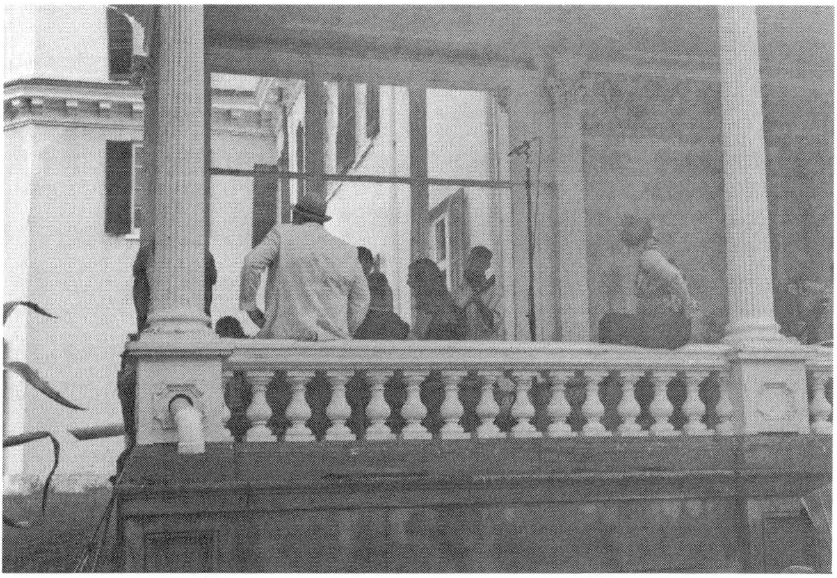

Figure 1.1 Ragnar Kjartansson: *The Visitors*, 2012 (The Porch). See also Plate 2.

Echo's Legacy

A piano whose sounds animate a nearby tuning fork. Metallic surfaces covered with sand that display wonderous figures when subjected to the movements of a violin bow. An aging engine causing your entire car to vibrate. Images that register the impact of strong magnetic fields on bodily organs. Prehistoric cave dwellers who use reverberant spaces to draw lines on rock formations. Modern-day musicians falling into each other's song and attuning their singing to the rhythms of a percussionist. A person agreeing with another person through empathy rather than rational argumentation. A poem whose words echo in my memory as if I had written or spoken them myself.

We have come to describe phenomena such as these, in all their difference, as examples of resonance: as instances in which something is affected by the vibrations, intensities, motions, or emotions of something else, as events in whose context one object causes another object to take on its own trembling. Inspired by Ragnar Kjartansson's *The Visitors* (2012), this book explores resonance as a model of art's fleeting promise to make us coexist with things strange and other. Approaching the sounds of installation art as a unique laboratory of hospitality amid inhospitable times, we will follow the echoes of distant, unexpected, and unheard sounds in twenty-first-century art to reflect on the attachments we pursue to sustain our lives and the walls we need to tear down to secure possible futures. As this book begins its journey, however, it is important first to discuss the often wildly different uses of the term "resonance" in past and present and complicate given assumptions, as they exist at least in everyday language, that resonance describes something unquestionably valuable or desirable in and of itself.

Consider the famous collapse of the Tacoma Narrows Bridge in 1940, triggered by peculiar wind conditions that caused the 2,800-foot construction to vertically oscillate beyond control. Physics textbooks typically present the collapse of the bridge as an example of forced vibration, the way in which one object or system—in this case, the wind—can force a neighboring object into large vibrational motion. Even though such transfers of energies may not produce any audible sound, physical engineers consider forced vibration as an instance of mechanical resonance, the term describing occurrences in which objects or systems respond to oscillating forces whose frequency is close to or identical with their own natural frequency.

No human was killed in the disaster at the Puget Sound. Its only victim was Tubby, a cocker spaniel owned by Leonard Coatsworth, the news editor of the *Tacoma News Tribune*. Coatsworth had been the last person to be on the bridge before its violent swaying brought it down. Here is his report of the events of November 7, 1940:

> Around me I could hear concrete cracking. I started back to the car to get the dog, but was thrown before I could reach it. The car itself began to slide from side to side on the roadway. I decided the bridge was breaking up and my only hope was to get back to shore. On hands and knees most of the time, I crawled 500 yards or more to the towers. . . . My breath was coming in gasps; my knees were raw and bleeding, my hands bruised and swollen from gripping the concrete curb. . . . Safely back at the toll plaza, I saw the bridge in its final collapse and saw my car plunge into the Narrows. With real tragedy, disaster and blasted dreams all around me, I believe that right at this minute what appalls me most is that within a few hours I must tell my daughter that her dog is dead, when I might have saved him.[1]

Though his experience was no doubt traumatic, Coatsworth understood how to retell the event engrossingly and dramatize his story. His report is rich in detail and texture, conjures compelling images to focalize his narration, seeks to capture the bridge's destructive oscillations with the rhythm of his syntax. Switching from past to present tense, he in the end effectively draws the reader even into his own feelings of fear, failure, and guilt. Literary criticism has various concepts at hand to analyze Coatsworth's strategies of emplotment, of appealing to the reader's affect and emotions. In everyday parlance, however, we have come to call Coatsworth's mode of telling the events of November 7 as an effort to make his experiences resonate with ours. He evokes striking images, memories, and emotions, not to offer a merely descriptive account of the event but to touch upon our affects and activate our empathy. To establish emotional resonance, for Coatsworth, is to invite strange readers into his story's space and share his harrowing memories in all their vibrancy.

The concept of resonance abounds with perplexing ambiguities and metaphysical subtleties, so many in fact that it appears virtually impossible to identify the concept's common denominator. At first, the term seems to belong firmly to the realm of audible sensations. "Resonare" in Latin originally meant the same as the equivalent in ancient Greek: "to resound" or "to echo." In Ovid's

[1] http://www.wsdot.wa.gov/tnbhistory/people/eyewitness.htm.

retelling of Echo's myth, Echo was of interest because of her unconditional but largely muted attachment to Narcissus, while the nymph's sad fate—the reduction of her voice to pure sonority, the shriveling of her skin, her final turning into stone—offered ample warnings not to replay what had caused Juno's initial wrath, namely Echo's boundless chatter. There can be little doubt about the masculinist framework in which Ovid's story has been told. Cursed to serve as a mere resonance chamber of the world, Echo's ability to resound the sounds of others certainly expressed the vitality, the utter attentiveness, of her emotions. Her inability to produce new sounds and utter words on her own, however, at the same time marked her downfall: unmitigated resonance could not but result in death.

The resonant life and death of Echo, a male fantasy of unique dimensions, strangely haunt today's uses of the term. Her being torn between the costs and blessings of living a life anchored in resonance has become our continued toggling between seemingly exclusive scientific and metaphorical, causal and affective, deterministic and non-deterministic, materialist and aesthetic understandings of the term. Mechanical engineers studying what aeroelastic flutter did in 1940 to the Tacoma Narrows Bridge use the term "resonance" to explain the physical world as a transparent dynamic of linear causes and effects, of measurable transactions and transductions, of identifiable stimuli and predictable responses. In the vocabulary of humanists as much as of everyday language, on the other hand, the concept of resonance serves a very different purpose: the term is typically used to describe the fuzzy landscapes of human emotions and affects, the complicated dramas of soul, body, mind, and empathy as they—like Coatsworth's retelling of the events of November 7—stretch out to, yet always fear not to touch exhaustingly enough onto, the lives of others. In stark contrast to scientists speaking about the physics of resonance, everyday situations resort to the concept whenever we cannot name exact causes and logical influences, whenever certain emotions, images, thoughts, or memories somehow chime with those of others, the term's presumed vagueness being its very strength, a reminder that the human might exceed logic, causation, and predictability.

Mechanical engineers claim Echo's resounding legacy as much as atomic particle scientists, molecular chemists, and sound technicians in order to name vibrational transfers between different objects and materials. For some, such transactions might be accompanied or even centrally concerned with audible sensations. For most, they aren't. More importantly, however, none of the scientific uses of the term seem to mesh with its metaphorical and affective understandings and the currency resonance has assumed in everyday language, in particular during the last decade or so. Whereas for some the concept offers a tool to soberly explain and analyze observable effects, for others resonance provides a lens of interpretation and speculation, a figure of thought meant to explore what transcends observation and analysis and cannot but remain inherently ambivalent. It is therefore tempting to suggest that C. P. Snow's famous 1959 distinction between the modern cultures of science and the humanities rips right through the concept of resonance as well,[2] with the term's exclusive applications documenting a profound absence of shared standards, methodologies, approaches, and values in different branches of contemporary knowledge production and society in general, between STEM and the aesthetic.

But wait. Your map of resonance might not be mine, yet even once divided cities and countries have managed to overcome cold war divisions and thus avoided repeating the narcissistic isolation and self-destruction of Echo's beloved. Does resonance know of some common ground beyond its discursive divisions? A ground unsettling Snow's two cultures? Neither mechanical engineers nor literary critics, so much is clear upfront, will disagree that resonance is a deeply relational concept—that it takes at least two different bodies, forces, materials, subjects, or objects to trade certain vibrations. So let me spin this a little further.

To resonate is to echo with something, is to enable or make an echo. It is to give an echo to something, to let something come and arrive, let something in and take place, so as to enable the kind of connection, attachment, and reciprocity we call resonance, whether it is strictly physical or bedazzlingly affective in nature. Cursed with, as Juno had put it, a voice much shorter than her tongue, Echo fell in love with and furtively followed Narcissus's footsteps.[3] Though unable to lend words to her passion, she certainly was neither numb nor dumb, neither void of agency nor closed to being touched by her immediate surrounds. Unlike her self-optimizing beloved, Echo's power to resound rested on her ability to get herself entangled and not to accept the borders, lines, distances, and differences that separate bodies, minds, objects, and subjects. She could only resound and resonate because she knew—stealthily or not—how to lean toward the other and thus allow this other to affect her.

[2] Charles Percy Snow, *The Two Cultures* (1959, Cambridge: Cambridge University Press, 1998).

[3] Ovid, *The Metamorphoses*, a new verse translation by Allen Mandelbaum (San Diego: Harcourt Brace & Company, 1993), 92: "When Echo saw / Narcissus roaming through the lonely fields, / she was inflamed with love, and—furtively— / she followed in his footsteps. As she drew / still closer, closer, so her longing grew / more keen, more hot—as sulfur, quick to burn, / smeared round a torch's top bursts into flame / when there are other fires close to it."

Echo's story has, for many good reasons, little currency in #MeToo times.[4] And yet, when read against the grain, Echo's fate communicates claims and insights that are worthy of being remembered and recuperated. Neither hearts nor bridges begin to tremble if they do not bring something to the table, if they do not open a door and thus offer a space to that which enters their perimeters and makes them vibrate. The analogue to Echo's unwanted, but nevertheless extraordinary, receptivity to the world is what mechanical engineers call the natural frequency of a given material or object, the inherent pulsation of all matter. Similar to how caves offer specific features and properties—affordances—to enable the wondrous effects of echoes, so do objects resonate with the vibration of other neighboring objects because they embody more than mere passive receptacles. Like good listeners listen out to the world around them, material objects, minds, and emotions resonate with others because they owe certain agential powers all of their own. They can take on the vibrations of other objects or forces because they have the capacity to affect the world themselves, to lean toward possible encounters and entanglements without necessarily pursuing any explicit purpose or strategic design.

Resonance, then, is both a strictly scientific and an unapologetically metaphorical concept. It is about measurability as much as it is charged with poetic vagueness. For some, it describes observable facts. For others, it provides a figure of thought, meant to evoke what exceeds observation. What all the word's different uses share, however, is to present resonant matter, the matter of resonance, as a condition in which one entity allows and welcomes another into its space and times without putting up fences and resistances, without expecting, anticipating, or generating some sense of reciprocity. Though resonance designates modes of attachment categorically clear to some and productively fuzzy to others, at heart it identifies relationships between dissimilar agents in which the strange, foreign, and other can interface with the known, familiar, or inherent across their very differences.

In the realm of human affairs, we call such openness hospitality: a condition requiring, following Derrida's reading of Kant, "that I open up my home and that I give not only to the foreigner (provided with a family name, with the social status of being a foreigner, etc.), but to the absolute, unknown, anonymous

4 For a seminal rereading of Echo's myth from a feminist perspective in the field of film studies, see for instance Amy Lawrence, *Echo and Narcissus: Women's Voice in Classical Hollywood Cinema* (Berkeley: University of California Press, 1991). See also Mary Noonan, *Echo's Voice: The Theaters of Sarraute, Duras, Cixous and Renaude* (New York: Taylor and Francis, 2014).

other, and that I *give place* to them, that I let them come, that I let them arrive, and take place in the place I offer them, without asking them either reciprocity (entering into a pact) or even their names."[5] Resonant matter, in all its different versions, is hospitable matter—matter that enacts or entertains processes of unconditional hospitality. It asks no questions, draws no pacts or agreements upfront, expects no tradeoffs for letting the other's vibratory energies come, arrive, and enter. Like a generous host, it simply takes on the vibrations of the other. We don't need to look at the Tacoma Narrows Bridge in order to remind us of the fact that gestures of absolute hospitality can often produce devastating effects. Yet the bridge's catastrophic fate offers a stark reminder that resonance, like absolute hospitality, is agnostic to what is given place; that we, in assessing the power of resonance as a structure of hospitable entanglement, should not naively think that it will automatically put forth the good and desirable; that resonant vibrations, even when at their most mechanic and deterministic, are fundamentally ambivalent in nature. Though at first seemingly inhabiting very different conceptual worlds, the two central terms of this book—resonance and hospitality—share common ground because both designate vibrant structures of attachment and entanglement while either of the two concepts draws our attention to the curious relationship of affect and effect, of emotional intensity and physical determination, of what is us and what isn't.

With Ragnar Kjartansson's *The Visitors* as its hub and anchor, *Resonant Matter* profiles different figurations of resonance and vibrant coexistence in contemporary art. It approaches the work of various artists as aesthetic laboratories of hospitality, of strange entanglements and entanglements of strangers; as events in which heard and unheard vibrations excite other matter and invite acts of resonant listening. It is said that sound and music have the enchanting capacity to touch upon its listeners at a distance. Listening exceeds the operations of the ear, the realm of acoustics. It is multisensorial, moving different sensory registers, being moved by subjects in motion. In this book, we proceed from the vibrational and resonant aspects of sound—its ambivalent hovering between science and aesthetics, fact and metaphor, causality and affect—to explore how sound in twenty-first-century art offers a home for the unconditional encounter with the strange, foreign, and other. At the book's center are projects that, in investigating both the scientific and affective registers

[5] Jacques Derrida and Anne Duforurmantelle, *Of Hospitality*, trans. Rachel Bowlby (Stanford: Stanford University Press, 2000), 25.

of resonance, welcome their listeners into wondrous spaces as much as they ask our ears and bodies to let unheard-of sounds to come in and give place to them in the home of our listening. A training ground for acts of unconditional hospitality, the vibrational aesthetics of sound art not only brushes against dominant distinctions between nature and culture, the human and the non-human, the physical and the metaphorical, matter and mind, the audible and the visual, the rational and the magic. It also asks tough questions about what it means to attend to sound in all its ambivalence—questions which amid the presence of ever-more mobile media and in face of neoliberal dynamics of singularization, self-management, and enclosure are deeply political in nature. To approach contemporary art through the conduit of resonance is to ask what it takes to live a life of entanglement and proximity today; to let what is strange arrive without expecting reciprocity or contractual payoffs. It is to ask how we can transform the fragmented spaces of our present into vibrant sites of multiplicity recognizing that, all things told, we all are nothing but visitors amid other visitors. Ovid's Echo died because Narcissus failed to find beauty and meaning in her resounding voice. In engaging with the aesthetics of resonance, sound art's promise today is to welcome Echo to assume her rightful place among the living and animate the spaces of their lives.

Between Magic and Science

In 1969 Alvin Lucier famously recorded *I Am Sitting in a Room* (Figure 1.2) first in his own faculty apartment and then at the Electronic Music Studio at Brandeis University. It enacted the very score that served as the piece's source material during the recording:

> I am sitting in a room different from the one you are in now. I am recording the sound of my speaking voice and I am going to play it back into the room again and again until the resonant frequencies of the room reinforce themselves so that any semblance of my speech, with perhaps the exception of rhythm, is destroyed. What you will hear, then, are the natural resonant frequencies of the room articulated by speech. I regard this activity not so much as a demonstration of a physical fact, but more as a way to smooth out any irregularities my speech might have.[6]

[6] Alvin Lucier, *I Am Sitting in a Room*, Lovely Music, Ltd., 1990, CD, album notes.

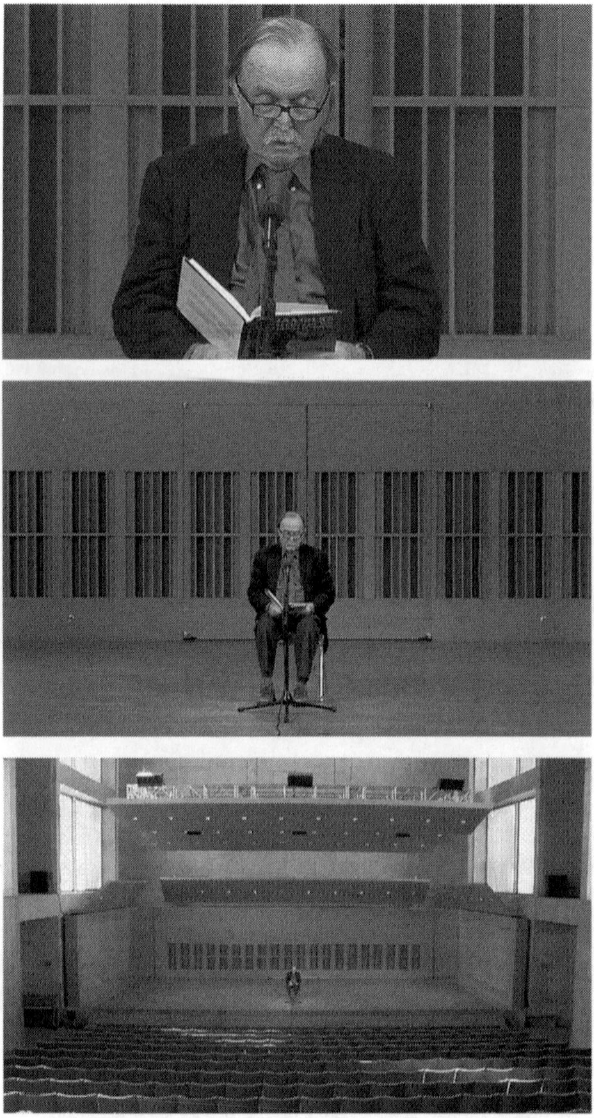

Figure 1.2 Alvin Lucier: *I Am Sitting in a Room,* 1969, 2010.

Following minimalist practice and the influence John Cage had on Lucier's artistic development circa 1970, *I Am Sitting in a Room* embraced process as content. The initial experiment relied on two Nagra tape recorders, a Beyer microphone, and a single amplified loudspeaker in order to record and re-record the very text that described Lucier's procedure. As it layered one feedback recording onto another, Lucier's speech became increasingly unintelligible while his voice allowed the room's resonant structure to come forth in its own right. Echoes typically need temporal sound differentials of 0.05 seconds and hence rooms larger than 10 meters in length in order to be perceived as such. In Lucier's intervention, the process of recording and re-recording the same words over and over again sidestepped these needs in such a way that the room's unique resonating frequencies eventually absorbed the lucidity of his meticulous explanations and the room's resounding echoes triumphed over the text's communicative content. Experimental composer Christoph Burns has described Lucier's performance as a conceptually rich and highly economical work "which questions the distinction between speech and music."[7] Lucier himself has recalled the piece as such: "Speech became music. It was magical."[8]

Lucier's stress on the enchanting qualities of resonance inadvertently expresses resonance's tough stance in modern cultures of enlightenment, rationalization, and criticality. As it stages a process by which resonant frequencies overwhelm the rational substrate of spoken text and thereby render human speech abstract, *I Am Sitting in a Room* praises as marvelous what troubled champions of modern rationalization in prior decades and centuries. Systematic research on the physics and physiology of resonance started as early as the seventeenth century with treatises such as *Du bruit* (1680) by French physician and architect Claude Perrault, trailing how René Descartes in his *Meditations* (1641) had already contemplated the human ear's abilities and limits to subtend his endorsement of unquestionable knowledge, rational sovereignty, and epistemological transparency. In the wake of Descartes and Perrault, enlightenment philosophers typically sought to pit reason against resonance, the first representing critical detachment, agency, culture, and freedom, the second embodying presence, proximity, touch, submission, nature, and necessity. As Veit Erlmann has shown in great detail, however, resonance and reason could never be separated as successfully as many a thinker desired.[9] One often ended up needing the other, both energized, grounded, and vindicated the operations of its presumed opposite. Rather than simply leading the subject into an abyss

[7] Christopher Burns, "Realizing Lucier and Stockhausen: Case Studies in Electroacoustic Performance Practice," https://pdfs.semanticscholar.org/65ab/. For extended discussions and appreciations of the piece's historical legacy, see, among many others, Douglas Kahn, "Alvin Lucier: *I Am Sitting in a Room*, Immersed and Propagated," *Oase* 78 (2009): 24–30.

[8] Alvin Lucier, *Music 109: Notes on Experimental Music* (Middletown, CT: Wesleyan University Press, 2012), 90.

[9] Veit Erlmann, *Reason and Resonance: A History of Modern Aurality* (New York: Zone Books, 2010).

of dark, muted, and merely affective interiority, resonance thus even during the heyday of enlightenment reason came to serve a double duty:

> On the one hand the concept of resonance, having been derived from core epistemological virtues such as intuition, observation, and experiment, names the natural mechanism governing the interaction of vibrating matter, such as strings, nerves, and air. As such resonance is the "Other" of the self-constituting Cartesian ego as it discovers the truth (of musical harmony, for instance) and reassures itself of its own existing as a thinking entity. On the other hand resonance names the very unity of body and mind that the cogitating ego must unthink before it uncovers the truth (of resonance, for example).[10]

In its very stress on the materiality of audition and the physical immediacy of vibratory transaction, resonance served and continues to serve as a potent reminder of the fragility of Western binary constructions of thought, truth, knowledge, and understanding. Similar to how categorical separations of nature and culture, in particular in our age of the Anthropocene, no longer appear viable, so we must think of resonance as an echo chamber, not a mere other, of practices, institutions, ideas, values, and hierarchies associated with the world of cognition. At the very least, as Erlmann himself concludes referring to the work of Bruno Latour, "resonance compels us to call into question the notion that the nature of things resides in their essence and that this essence can be exhausted by a sign, a discourse, or logos."[11]

It is the research of nineteenth-century scientists such as Hermann von Helmholtz that manifested this reciprocity of reason and resonance—the logic of resonant matter and the resonant qualities of scientific rationality—most exhaustingly.[12] A scholar for whom art and science mutually inflected each other, Helmholtz's work on acoustics is perhaps best known for its piano theory of hearing. It considered the ear's nerve fibers as analogous to the strings of a piano, and it claimed that these fibers resonated with incoming sounds similar to how piano strings tuned to certain frequencies would produce specific

[10] Veit Erlmann, "Resonance," *Keywords in Sound*, eds. David Novak and Matt Sakakeeny (Durham: Duke University Press, 2015), 177.

[11] Erlmann, "Resonance" 177.

[12] Though sound and music concerned Helmholtz throughout his career, his most influential research and insights were gathered in the 1862 publication, *On the Sensations of Tone as a Physiological Basis for the Theory of Music*, trans. Alexander J. Ellis (New York: Dover Publications, 1954). For more on Helmholtz and his place in nineteenth-century science and acoustics, see Benjamin Steege, *Helmholtz and the Modern Listener* (Cambridge: Cambridge University Press, 2012); and David Cahan, *Helmholtz: A Life in Science* (Chicago: University of Chicago Press, 2017).

tones and their partial overtones when activated through external stimuli. For Helmholtz, the relation between musical instruments and human organs was by no means meant to be merely metaphorical. The vibrational operations of a piano instead at once emulated and helped explain the operations of the human ear; technological, physiological, and aesthetic dimensions were to complement each other to inform the design of Helmholtz's experimental setups and generate measurable knowledge about how sound and music affected human audition. As he in this way envisioned the human ear as a resounding set of strings that transduced sound's vibrations to the human brain, Helmholtz situated resonance as a crucial medium to investigate both the empirical work of human attention and the structural shapes of compositional work, the aesthetics of music.

This returns us to Lucier's *I Am Sitting in a Room*, to its curious ways of exposing resonance to strategies of inversion and transposition. Rooms are meant to give place to intimacy, seclusion, and control. They protect subjects from contingencies and help vulnerable egos cultivate their sense of sovereignty. Though they often fail to live up to it, rooms come with the promise of contemplative interiority, a sense not only of focus and concentration but also of familiarity and homeliness, unattainable in the world beyond their perimeters. Lucier's play with resonance renders the familiar strange, the homely unhomely or uncanny, the interior exterior, the intimate wondrous, and he precisely thereby unsettles whatever we may take for granted about the attentional orders of the day. Instead of listening to a voice safely protected by the walls of a room, the listener of *I Am Sitting in a Room* by the end of the piece hears no more and no less than the sonority of the room itself, its resonant frequencies. Instead of featuring the voice as a tool defining human autonomy and sovereignty, Lucier explores resonance as a medium folding object and subject into one single dynamic. Instead of presenting a room as a safe haven for reasoning and communication, *I Am Sitting in a Room* approaches spatial infrastructures as acoustic ecologies, as sites in which the border between the familiar and the foreign, system and environment, logos and affect, are everything but stable and secure. In this way, Lucier poses the question of resonance as a question, not simply of how listeners attend to sounding objects, but of what and who this attentive listener might be in the first place.

Lucier's posing of the question of attentive listening is a question that—as we will see in our engagement with Kjartansson's *The Visitors*, over and over again—matters profoundly in our media-saturated and attentionally challenged twenty-first century. Well-meaning nostalgia for a past of concentrated listening, viewing, or reading no longer helps to navigate the present and future. The twenty-first century's world, writes Malcom McCullough, has been "filling with many new kinds of ambient interfaces. Nothing may be designed on the assumption that it will be noticed. Many more things must be designed and used with the ambient in mind. Under these circumstances, you might want to rethink attention."[13] To think about the power of resonance today is to rethink how we inhabit the landscapes of the present. It is to explore new modes and understandings of what it means to attend to and be absorbed by the overwhelming number of stimuli that have come to surround us almost everywhere and at all times. Regardless of whether we, like Lucier, sit in rooms of our own choosing or move restlessly through strange urban spaces and their mediascapes, to reflect on the power of resonance today is to reflect on different logics of attention and entanglement, on what might attach us to both the proximate and the distant.

The Dialectics of Vibration

It is commonplace to say that everything vibrates and that what we call sound only opens a small window onto the vibratory energies of all matter. We can feel or see the vibratory qualities of audible sound at certain frequencies even if we, like Homer's Odysseus, sealed our ears with wax: think of the rumbling deep basses create in your stomach. But we can't hear a much larger range of vibrations no matter how hard we try because the tympanic membranes of our ears—in spite of their impressive engineering—provide rather limited media to deliver the cyclic compression of air caused by the vibration of matter below 20 Hz or above 20,000 Hz. The vibratory energies of infrasound and ultrasound exist outside the window of human audition. What Steve Goodman calls "unsound" constitutes the world's acoustical unconscious—if, and only if, we think of this world as something that pivots around human history and perception.[14] Or it serves as a stark reminder to recognize the limits of the human, the way in which the human shares time and space with inanimate objects and animated events that

[13] Malcolm McCullough, *Ambient Commons: Attention in the Age of Embodied Information* (Cambridge, MA: MIT Press, 2013), Kindle file.

[14] Steve Goodman, *Sonic Warfare: Sound, Affect, and the Ecology of Fear* (Cambridge, MA: MIT Press, 2010), 189–95.

exceed the grasp of individual self-management and categorically question the possibility of, as Salomé Voegelin has put it, anthropocen(tr)ic meta-positions.[15]

In his research on paleolithic art, scholars such as David Lewis-Williams have associated the production of early cave painting with the resonant vibrations of low-frequency drum-beats, their ability to induce trance-like states and enable out-of-body experiences. Scientific findings suggest, Lewis-Williams explains, that resonant areas in caves such as the one at Chauvet in Southern France "are more likely to have images than non-resonant ones. The implication is that people performed rituals involving drumming and chanting in the acoustically best areas and then followed up these activities by making images."[16] Resonant vibrations—the effects of rhythmic drumming, the echoes of invisible animal sounds, and the reverb of unexplainable topographical and geological noises—in Lewis-Williams's view marked the origin of what much later generations would call art. As he concludes,

> The factors that governed the placing of images were far more complex than resonance alone, and included the way in which the topography of a cave was conceived and the locations where specific kinds of spiritual experiences were encouraged. Still, we need not doubt that resonance and echoes added to the effect of subterranean Upper Paleolithic rituals. The caves, if not the hills, were *alive* with the sound of music.[17]

As it explores the nexus between space, music, ritual, art, and spirituality, recent research in fields such as aural architecture traces the many echoes paleolithic uses of resonance have found in much later ages in order to make different places vibrate and speak, whether we think of classical Greek open-air theaters, the design of Gothic cathedrals, or the more recent setup of immersive

[15] "Hearing is full of doubt: phenomenological doubt of the listener about the heard and himself hearing it. Hearing does not offer a meta-position; there is no place where I am not simultaneous with the heard. However far its source, the sound sits in my ear. I cannot hear it if I am not immersed in its auditory object, which is not its source but sound as sound itself. Consequently, a philosophy of sound art must have at its core the principle of sharing time and space with the object or event under consideration." Salomé Voegelin, *Listening to Noise and Silence: Towards a Philosophy of Sound Art* (New York: Continuum, 2010), xii.

[16] David Lewis-Williams, *The Mind in the Cave: Consciousness and the Origins of Art* (London: Thames and Hudson, 2002), 225.

[17] Lewis-Williams, *The Mind in the Cave*, 226. For related overviews of what inspired early cave art, and whether we should consider it as art in the first place, see, among many others, Max Raphael, *Prehistoric Cave Paintings*, trans. Norbert Guterman (New York: Pantheon Books, 1946); Ellen Dissanayake, *Homo Aestheticus: Where Art Comes From and Why* (Seattle: University of Washington Press, 1995); Jean Clottes, *Cave Art* (London: Phaidon, 2008); and David S. Whitley, *Cave Paintings and the Human Spirit: The Origin of Creativity and Belief* (Amherst: Prometheus Books, 2009).

surround sound entertainment systems.[18] Bissera Pentcheva's work on the resonant acoustics of the Hagia Sophia, built in 537 CE in what is now Istanbul, is one among many fascinating studies investigating the complementary ways in which space and sound, location and reverberation animated worshippers in the Byzantium and created aural experiences of transcendence and quasi-mystical enchantment.[19] In the world of contemporary compositional practice, Lewis-Williams's animated caves have found their most prominent resurgence in the work of Pauline Oliveros. Since the late 1980s, her project of deep listening probed the resonant acoustics of underground cisterns, cathedrals, and caves, not only to produce new sounds, but in so doing to change dominant structures of attentive listening, sonic awareness, and gendered aurality.[20] For Oliveros, to perform and record song, music, and sound in reverberant spaces was to investigate acoustic ecologies whose vibrational energies—reminiscent of paleolithic rituals—blurred preexisting boundaries between the human and the nonhuman, the materiality of space and the creativity of consciousness.

Vibration, writes Shelley Trower, "is not itself a material object at all, but it is bound up with materiality: vibration moves material, and moves through material. Rocks are among the most solid, the most palpable and stable of objects, yet in an earthquake they are of course subject to vibratory movements."[21] Like resonance, which owes its existence to the movements of vibratory forces, vibration unsettles our view of things according to opposing principles of subject and object, inanimate and animate, active and passive. Similar to sound, vibration is an inherently relational phenomenon, that is, it only exists because different entities touch upon and agitate each other. Because vibration dynamically connects, entangles, and attaches different bodies or forms of matter, its study cuts across and includes diverse fields of knowledge such as physics, neurology, psychology, aesthetics, political science, and metaphysics. It therefore should come as no surprise that vibration's reputation among cultural critics is, to say the least, charged with ambivalence. Because it may lack proper object status, the vibratory in itself may be neither good nor bad, progressive nor

[18] See, for instance, the fascinating contributions to Barry Blesser and Linda Ruth-Salter, eds., *Spaces Speak, Are You Listening? Experiencing Aural Architecture* (Cambridge, MA: MIT Press, 2006).

[19] Bissera V. Pentcheva, *Hagia Sophia: Sound, Space, and Spirit in Byzantium* (University Park, PA: Penn State University Press, 2017).

[20] Pauline Oliveros, *Deep Listening: A Composer's Sound Practice* (Lincoln, NE: iUniverse, 2005). See also, Martha Mockus, *Sounding Out: Pauline Oliveros and Lesbian Musicality* (New York: Routledge, 2007).

[21] Shelley Trower, *Vibration: A History of the Pleasure and Pain of Sound* (New York: Continuum, 2012), 6.

regressive, pleasurable nor painful. For some, however, vibration's association with materiality and affect, its privileging of matter over reason, poses severe risks for the integrity of human affairs, the structured regulation of social interaction, and the sustainability of democratic public spheres. Pluralist commons, they argue, require restraint and rational deliberation, whereas any effort to release vibratory energies into the political cannot but approach the anti-pluralistic drive of modern populism, demagogism, totalitarianism, and perhaps even fascism. For others, vibrational fields and energies provide resources of individual healing and therapeutic recovery, supporting fatigued and traumatized modern subjects with what is necessary to act as autonomous, ethically mindful, and accountable citizens. Following Hans Jenny's notion of cymatics,[22] a whole branch of contemporary medicine uses therapeutic tools such as musical rhythms, high-intensity focused ultrasounds, tuning forks, and vocal expressions either to recalibrate the body's denser physical energy fields; or to directly affect the health of organic tissues and even with remarkable success to cure the damage of cancer cells.

The first position—vibration's association with subjection, heteronomy, and even terror—has been perhaps most vividly explored in Mark Bain's *The Live Room: Transducing Resonant Architecture*, first installed at V2_Lab for the Unstable Media in Rotterdam, the Netherlands, in 1998 (Figure 1.3). A practitioner whose work over the last decades has consistently focused on the interactions of audible sound, architectural spaces, vibrational oscillations, and infrasonics, Bain in *The Live Room* relied on a number of vibrating devices mounted directly into the walls and floors of the exhibition space. Their pulses activated the building's resonant frequencies and thereby transformed the entire structure into a resounding speaker. Though certainly aware that subsonic sounds and resonant vibrations can create soothing effects, Bain's *The Live Room* instead aspired to tune the visitors' senses to vibratory movements whose purpose was to unsettle our perception and navigation of built space. As if being hit by a prolonged earthquake, the viewer was expected to sense extreme discomfort, disorientation, and fear as the room's rumbling seemed to revoke any difference between interior and exterior realms, spaces of perception and perceived space. A later review listed nausea, headaches, gag reflexes, and urges to defecate as immediate effects, caused by how *The Live Room*'s vibrations messed with the

[22] Hans Jenny, *Cymatics: A Study of Wave Phenomena and Vibration* (Eliot, ME: MACROmedia Publishing, 2001).

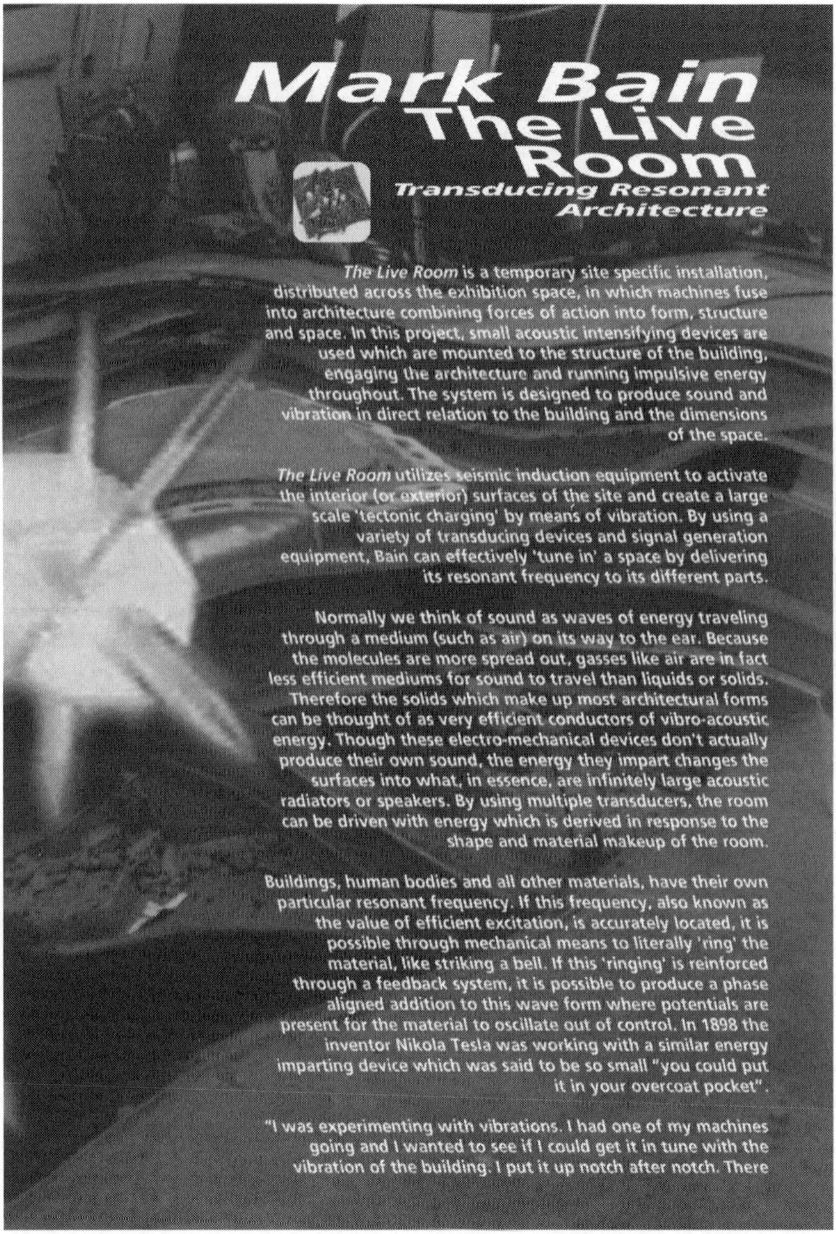

Mark Bain
The Live Room
Transducing Resonant Architecture

The Live Room is a temporary site specific installation, distributed across the exhibition space, in which machines fuse into architecture combining forces of action into form, structure and space. In this project, small acoustic intensifying devices are used which are mounted to the structure of the building, engaging the architecture and running impulsive energy throughout. The system is designed to produce sound and vibration in direct relation to the building and the dimensions of the space.

The Live Room utilizes seismic induction equipment to activate the interior (or exterior) surfaces of the site and create a large scale 'tectonic charging' by means of vibration. By using a variety of transducing devices and signal generation equipment, Bain can effectively 'tune in' a space by delivering its resonant frequency to its different parts.

Normally we think of sound as waves of energy traveling through a medium (such as air) on its way to the ear. Because the molecules are more spread out, gasses like air are in fact less efficient mediums for sound to travel than liquids or solids. Therefore the solids which make up most architectural forms can be thought of as very efficient conductors of vibro-acoustic energy. Though these electro-mechanical devices don't actually produce their own sound, the energy they impart changes the surfaces into what, in essence, are infinitely large acoustic radiators or speakers. By using multiple transducers, the room can be driven with energy which is derived in response to the shape and material makeup of the room.

Buildings, human bodies and all other materials, have their own particular resonant frequency. If this frequency, also known as the value of efficient excitation, is accurately located, it is possible through mechanical means to literally 'ring' the material, like striking a bell. If this 'ringing' is reinforced through a feedback system, it is possible to produce a phase aligned addition to this wave form where potentials are present for the material to oscillate out of control. In 1898 the inventor Nikola Tesla was working with a similar energy imparting device which was said to be so small "you could put it in your overcoat pocket".

"I was experimenting with vibrations. I had one of my machines going and I wanted to see if I could get it in tune with the vibration of the building. I put it up notch after notch. There

Figure 1.3 Mark Bain: *The Live Room: Transducing Resonant Architecture*, 1998.

visitors' vestibular systems, disturbed their sense of balance, and rendered any view of stable architectural planes and horizons impossible.[23] Using sonic and subsonic vibrations to shake visitors out of common feelings of certainty and boundedness, resonance here annulled the boundaries between the animate and the inanimate and as a result firmly incorporated the subject into the object. Bain's installation invited visitors to probe vibratory conditions that take hold of their entire bodies and no longer allow for any sense of control, autonomy, and self-direction; it enabled art patrons to experience frightful states of utter subjection in which experience itself is being evacuated.

Consider fuse*'s *Dökk* (2017) (Figure 1.4), a multimedia dance and video installation, as representing the other end of the spectrum: a vibrational exercise in whose context subjectivity becomes cosmic and absorbs the objective world in its entirety. Played out across ten consecutive stations or so-called rooms, the dancer's dance takes place in front of a semi-transparent holographic screen whose projected images at times show the universe at a scale of 100 million light years, at others at the scale of the intricacies of human neuronal networks. Six acoustical "ghost" tracks accompany the performance, each pegged to basic emotional states and activated in response to algorithms that carry out sentiment analyses of messages circulated through social media worldwide in real time. The common thread of *Dökk*, the Icelandic word for darkness, is interdependence at the highest level of complexity. Eighteen motion sensors are attached to the dancer's hands and feet, tracking her movements throughout the performance to affect the shapes and rhythms of the holographic images. The dancer's heartbeat is made audible via Bluetooth devices for the audience. These beats punctuate the ghost tracks and fold the show's interstellar imagery into what are the most intimate human sounds. Two of the ten team members of the Italian-based cooperative of fuse*, Mattia Carretti and Riccardo Bazzoni, describe *Dökk*'s exploration of seamless vibratory interaction and resonance as follows:

> All the things that surround us are nothing but a collection of atoms, particles, and electromagnetic fields, vibrating without any apparent meaning. When these impulses are interpreted by our mind, they become colors, tastes, music, memories, and emotions—the foundation of what each one of us perceives as reality. Everything that happens in *Dökk* is represented with a universe that is

[23] Jo-Anne Green, "Mark Bain: Works X 2," *Networked_Music_Review* (September 30, 2007), http://turbulence.org/networked_music_review/2007/09/30/mark-bain-works-x-2/?author=4.

> initially displayed with its real shape with particles representing the galaxies positioned exactly in real coordinates. Then, during the journey, this universe changes its shape assuming different configurations until it reaches a new balance at the end of the show.[24]

With its scenography designed to deliver a deep sense of resonant connectivity, *Dökk* recast the world as a space of boundless interiority. The show invites its

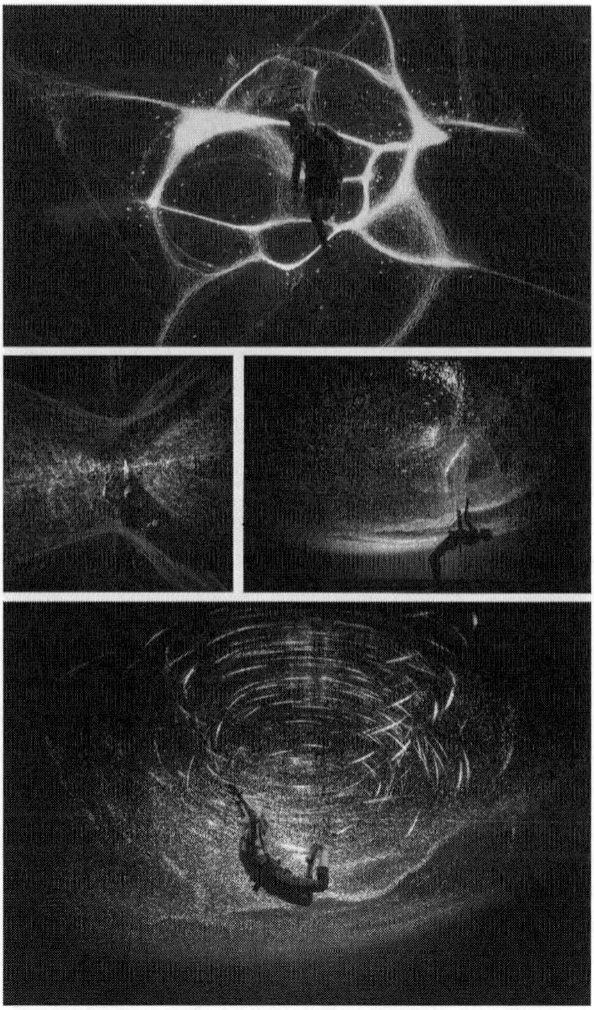

Figure 1.4 fuse*: *Dökk*, 2017.

[24] https://cycling74.com/articles/artist-focus-fuse*.

audience to a vibrant journey through different digital environments from the galactic to the spaces of the unconscious. Because, in fuse*'s understanding, molecules vibrate as much as entire universes, each can allegorize and model the other, so much so that the dancer's ecstatic moves on stage hold no less than the power to animate the entirety of the known and unknown world. In fuse*'s at once spectacular and metaphysical conception, vibration—as channeled through advanced digital technology—figures as an elemental medium. It not only interconnects the smallest and the largest but serves as a trigger to reenchant a seemingly disenchanted modern world and recast our environments as ones in which traditional distinctions between subjectivity and objectivity no longer appear possible or necessary. Everything in *Dökk*'s vibrating universe resonates with the energies of subjectivity. As a result, little will be left for the cosmically expanded subject to experience as strange and other. Radiating triumphant, total resonance here in the end possibly eclipses experience and mutes itself.

The resonant agendas of Bain's architectural structure and fuse*'s ten networked rooms could not be more different in nature. The former uses vibratory forces to engineer a world voided of human subjects and interiority, while in the latter vibrational movements revoke the exteriority of the world and animate everything with the élan of subjectivity. In the first vibration is meant to profile the power of cold totalitarian control; the second taps into the vibratory to stage hallucinogenic sensations of boundlessness, flow, and immersion. And yet, what both projects share in spite of their difference is the fact that both, in folding the strange and foreign into radically expanded formations of either the objective or the subjective, erase the very conditions for gestures of hospitality. Neither *Live Room* nor *Dökk* offers rooms that can give place to unknown or anonymous others and that let them arrive without asking them to give up their otherness and to enter pacts expecting seamless assimilation or integration.

As these two examples suggest, the question about the politics of resonance and hospitality, of resonance as a mode of hospitality, is intimately tied to the question of immersion, of how sounds and vibrations are able to envelop and absorb our registers of sensory awareness, situate us in the here and now as if there was no mediation, and possibly subject subjects to uncontrollable stimuli of their environments. Sounds and vibrations, Frances Dyson writes, surround:

> Immersed in sound, the subject loses itself, and, in many ways, loses its sense. Because hearing is not a discrete sense, to hear is also to be touched, both physically and emotionally. We feel low sound vibrate in our stomachs and start to panic, sharp sudden sound makes us flinch involuntarily, a high-pitched

> scream is emotionally wrenching: sound has immediate and obvious physical effects. In listening, one is engaged in a synergy with the world and the senses, a hearing/touch that is the essence of what we mean by gut reaction—a response that is simultaneously physiological and psychological, body and mind.[25]

The vibrational force of sound, Dyson continues, has atmospheric qualities. It acts on our perception like dense fog, engulfing our senses and nestling below our skin, yet seemingly disappearing and eluding words and discourse when we seek to approach it conceptually. Sound's resistance to theory and theorization, however, should not cause us to obscure or mystify the technological and cultural origins of audio media, however invisible, ephemeral, intangible, or un-mediated they often seem. Even the most effusive vibrations rely on certain rhetorics; neither audible nor inaudible sounds exist without media—technologies in a broad sense of the word—that transport and refract vibrational energies.

The immersive effect of vibrational energies can facilitate, as Goodman argues, a transduction of the tensions and conflicts of modern life and transform entrenched structures of fear, dread, and enmity. It can model new types of interactions and communities. Audible or not, vibration and its media organize space and time, and in this way they can not only provide abstract models of the social order but also actively create and sustain the rhythms and structures of banding and bonding,[26] solidarity and resistance. Needless to say, however, such vibratory movements can also serve as funnels of colonization, control, and violence: forces that in connecting separate entities, be they organic or inorganic, small or cosmic, weaponize affect and, by tampering with tonalities and temperaments, launch wars with other means even prior to any explicit ideological confrontation.[27] Whether vibration underwrites colonizing gestures of incorporation or enables the nexus we call hospitality is never a given. It is an open question, an aesthetic as much as a political issue, in need of careful deliberation and analysis. Vibration precedes meaning and discourse, but its meaning itself cannot be assessed without discourse. *Resonant Matter* approaches *The Visitors* and other recent works by sound, media, and installation artists as

[25] Frances Dyson, *Sounding New Media: Immersion and Embodiment in the Arts and Culture* (Berkeley: University of California Press, 2009), 4. See also Dyson's *The Tone of Our Times: Sound, Sense, Economy, and Ecology* (Cambridge, MA: MIT Press, 2014).

[26] Mark Slobin, "Ensembles—Banding versus Bonding," *Subcultural Sounds: Micromusics of the West* (Middletown, CT: Wesleyan University Press, 2000), 98–108. See also Michael Denning, *Noise Uprising: The Audiopolitics of a World Musical Revolution* (London: Verso, 2015), Kindle file, loc. 266.

[27] Goodman, *Sonic Warfare*, xii–xx.

places that invite such reflections about the aesthetics and politics of vibration. Its principle task is to discuss what it takes to understand resonant vibrations, in their ambivalent location between the scientific and the metaphorical, as sites of hospitality, very well aware of the fact that vibrational forces have the very power to eliminate what makes hospitality possible in the first place.

Sound (in) Art / Sound(,) Art

Sound art is a beast with many faces. There is no principal agreement on whether the term designates a movement, a genre, or a medium of artistic practice, let alone on how to distinguish it from experimental music, on the one hand, and performance and installation art, on the other. The term, popularized in the 1990s, has generated a good number of major exhibitions over the last decades, has gained considerable traction in the twenty-first century to fund grants for practitioners, residencies, art centers, educational and publication programs, and specialized tracks in select art and music schools across the globe.[28] And yet, no one is really happy with the concept of sound art. While the perhaps somewhat opportunistic comma this book's subtitle places between the words "sound" and "art" recognizes the conceptual instability of the term "sound art," it also approaches the term's messiness as a source of insight, as a place of unsettling questions rather than unbending identifications. The subtitle's punctuation mark is meant to offer a pause that in its very gesture of separating terms urges the reader to explore possible connections and allegiances: to think of sound art as a project and disruptive mode of artistic practice rather than a tool of marketable classification.

In his 2007 landmark compendium, Alan Licht defined sound art as a rather hybrid form of artistic practice that "rejects music's potential to compete with other time-based and narrative-driven art forms and addresses a basic human craving for sound."[29] For the purpose of his study, he then specified the domains of sound art along the lines of three different categories: (1) as a sound environment installed in space and set to be exhibited like a visual

[28] For more on the surge and challenges of sound art exhibitions in contemporary museum and gallery spaces, see Seth Cluett, "Ephemeral, Immersive, Invasive: Sound as Curatorial Theme, 1966–2013," *The Multisensory Museum: Cross-Disciplinary Perspectives on Touch, Sound, Smell, Memory, and Space*, eds. Nina Levent and Alvaro Pascual-Leone (Lanham, MD: Rowman & Littlefield, 2014), 109–18.

[29] Alan Licht, *Sound Art: Beyond Music, Between Categories* (New York: Rizzoli, 2007), 16.

artwork, (2) as a visual artwork that in one way or another includes sound-producing functions, and (3) as the sonic extension of an artist's particular aesthetic that is typically carried out in other mediums. Barbara London, when curating and conceptualizing MoMA's widely discussed 2013 show *Soundings: A Contemporary Score*, considered sound as an ever-more important medium in contemporary art investigating the connections between art, music, and noise in objects, installations, and performances while placing pressure onto the hegemonic institutions of the art world. In London's understanding, sound art, including work addressing silence and non-cochlear vibrations, expands and diversifies the place of art in contemporary society, not least of all because it—in spite of its growing canonization—continues to question "how and what we hear, and what we make of it."[30] More recently, Christoph Cox considers the term of sound art "clumsy and inexact, lumping together heterogenous art works and practices, ignoring their differences and particularities and obscuring their connections and allegiances to other artistic fields and categories."[31] And yet, in all its thorniness the term appears useful to him, not only because it registers the extent to which sound has gained considerable importance to various artistic practices, but also because the term invites questions about the ontology and metaphysics of sound, art, and hearing largely rendered mute by twentieth-century analytic and continental philosophy. Sound art, for Cox, reopens the very case of materialism, flux, presence, and vitalism critical theory had sought to shut down throughout the last century.[32]

The concept of sound art might represent a beast with many faces, but there are of course many other terms we use frequently in spite of the fact that their meanings are wide open to interpretation and perspective. *Resonant Matter* holds on to it as an expedient short for "sound (in) art," for "sound(,) art," precisely because the term in all its ambiguity performs what distinguishes the core of audition as well: it expresses an approach rather than an unequivocal fact, its heterogeneity undergirds curiosity rather than hairsplitting gestures of ordering. This decision may at first raise additional eyebrows in particular among readers who quite rightly consider works such as Kjartansson's *The*

[30] Barbara London, "Soundings: From the 1960s to the Present," *Soundings: A Contemporary Score* (New York: The Museum of Modern Art, 2013), 10.

[31] Christoph Cox, *Sonic Flux: Sound, Art, and Metaphysics* (Chicago: University of Chicago Press, 2018), 4.

[32] See also Douglas Kahn, "Sound Art, Art, Music," https://soundartarchive.net/articles/kahn-2006-Sound%20Art,%20Art,%20Music.pdf; and Andreas Egström and Åsa Stjerna, "Sound Art or *Klangkunst*? A Reading of the German and English Literature on Sound Art," *Organised Sound* 14.1 (2009): 11–18.

Visitors as screen-based installation art and many of the other pieces discussed in these pages as examples of performance art or of experimental music. Why not simply speak of *The Visitors* as a work of multichannel video art? What do we gain when confronting Kjartansson's or other installations with the concept of sound art? Let me answer these questions by discussing why and how *Resonant Matter* understands the term of sound art—sound (in) art; sound(,) art—as a question in the first place, as an unruly node of artistic practices and experiences that draws on sound to probe the locations, extensions, affects, and claims of various artistic media and genres today.

I begin answering these questions by naming the perhaps obvious: whatever is called sound art today continues and in many respects amplifies the impact of video on gallery spaces and institutional valuations of art and art music since the mid-1960s, in particular the effect of experimental work mixing moving images, sculpture, music, and sound on both modernist conceptions of the museum as a neutral container of aesthetic appreciation and traditional ideas of art music as a genre best appreciated in hushed auditoriums and in states of physical immobility. Though video featuring sound and music is well-respected and in fact ubiquitous today, its expansion during the 1960s and 1970s—think of the jarring efforts to visualize music by early video practitioners such as Nam June Paik, Charlotte Moorman, Steina Vasulka, and Tony Conrad—at first certainly did not meet hospitable hosts amid established institutions of art and art music alike. As Holly Rogers explains:

> First, a work could activate its surroundings by incorporating both performance space and the people within it into its composition. Second, the use of moving images placed the element of time into a space conventionally filled with static objects and, in so doing, presented the exhibition-goer with a dimension usually reserved for the concert hall or theatre. Finally, and yet most radically, video, as an audiovisual medium, introduced sound and music into the gallery environment, a space normally occupied by silent works.[33]

And as Rogers concludes: "Expanding into the gallery space spatially, temporally, and aurally, then, video works required a radical re-evaluation of art exhibition practice (in terms of curation, preservation, funding, and patters of audience engagement) and, to take one step further, the defining parameters of art itself."[34]

[33] Holly Rogers, *Sounding the Gallery: Video and the Rise of Art Music* (Oxford: Oxford University Press, 2013), Kindle file.

[34] Rogers, *Sounding the Gallery*, Kindle file.

Though twenty-first century art museums are almost unthinkable without the presence of moving images and even concert halls and opera stages today include video as part of their choreographies, the acoustical dimensions of video presentations remain somewhat challenging and remind us of how video and early art music video provoked established norms and hierarchies.

When tasked to install art with sound or video-based art music within existing gallery spaces, curators today continue to struggle with these earlier provocations and resistances. Sound is promiscuous and often indiscreet. It inflects and infects objects of artistic expression and easily invades spaces designed for the reception of other works. Like sound waves themselves, sound in art has little respect for different strategies of containment and classificatory identification. It activates its surroundings and temporalizes seemingly static spaces. What this book features as sound art, however, neither makes any excuses for such promiscuousness, nor does it uphold the way in which previous centuries had sought to separate the spaces for visual art and art music, stillness and mobility, spectatorial distance and participation. Sound art instead embraces the radiant qualities of sound waves to shape space beyond normative frameworks of focused attention, immobility, and disembodied appreciation, and we therefore do well to consider it, not as a self-contained medium or genre but as a variable set of aesthetic strategies that produce and modulate objects, spaces, and practices of aesthetic experience. Similar to how film noir in the 1940s and 1950s wandered across different film genres and thereby raised complex questions about the present and future of cinema, sound art refolds the genres and media of contemporary art in order to brush against existing domains of the aesthetic and explore how different media of art affect their audiences. Understood in this sense, sound art first and foremost treats vibrations, heard and unheard ones, as sculptural elements. It offers affordances to experience sound, not as a pure object of auditory pleasure, but as being mediated through multiple senses, including our senses of touch and motion and our awareness of the borders and extensions of our bodies in space. Sound in art interrogates the where, why, and how of contemporary listening. As it renders audition as something we can never take for granted, it invites audiences to encounter sound as a medium of world-making, as a force that by affecting various registers of perception retunes how we navigate space and place to begin with.

Mediation matters to sound art, and so it comes as no surprise that sound art will often work with screens and moving images, even multiscreen choreographies, to explore the nexus between the auditory and the visual and to supplement, deflect, or complicate architectures and ecologies of perception. As a mode of artistic practice, sound art allows works of screen-based installation art to ask new questions about the modalities of spectatorship, about how we attend to mediated sounds and images amid contemporary conditions of ever-more ambient, ubiquitous, and mobile media uses. Just as importantly, sound art tends to question prevalent notions of authorships and the self-sufficient "work" it may create. Experimenting with the seemingly intangible and ephemeral dimensions of the acoustical or vibrational, sound art generates decisively weak objects, unstable and transient events and relationships whose meaning may shift profoundly according to context. Though some maker is likely to operate in the background, sound art may very well exist without the physical presence of a performer—a performer's body whose actions, movements, and gestures, whether spontaneous or highly choreographed, address the audience's senses within the perimeters of a particular setting.

There are many ways to tame the beast called sound art, identify its objects and extensions, and define its difference and specificity. None of these can expect consensual support among practitioners and critics alike and many, in fact, might draw scorn for their very effort to dissect the field of artistic production and thereby, presumably, discipline artistic creativity. *Resonant Matter* hopes to avoid such suspicions. What I understand as the mode of sound art does not seek to relabel work typically considered as experimental music, screen-based installation work, or performance art. Instead, it simply wants to describe modes of artistic practice that work with sounds and vibrations, echoes and resonances in order to rethink the relations of music, installation, video, and performance and in this way question ontological juxtapositions of genres or media of art in the first place. To practice or attend to sound art is to pause and attune our senses to the aesthetics of resonant vibrations, no matter whether a "work" in questions formally belongs, or is intended to belong, to the realm of experimental music, video, installation, or performance art. It is not to stick classifying tags to certain artistic practices, but to disrupt such labeling and invite us attentively to listen to what the eye does not see, to hear what escapes the ear, and to touch or be touched by what resists the disciplining power of seemingly self-reliant genres and media.

What this book understands as sound art—sound in art, sound(,) art—at heart concerns itself with the extent to which vibrating matter matters to various forms of art because it incites or is welcomed by other materials to vibrate. Interested in the interconnectedness of materiality and meaning, the human and the nonhuman, sound art recognizes different figurations of resonance as the driving forces behind a work's, an installation's, a song's, or a performance's formal shapes and aesthetic energies. Echo's life and legacy is therefore much more than a mere side concern of what I consider the leading question of sound art—sound art's questioning of art. For Echo's (unchosen) talent in resonant listening continues to figure as sound art's indispensable resource and her painful marginalization in extended (his)stories of visual mastery and control remains sound art's critical point of contention, of aesthetic intervention and political reversal. Whatever its hosting medium or genre, sound art transforms Echo's destiny, her blessing and doom to resound the sounds of others, into creative opportunities and counterfactual narratives. Her openness to the world's vibrant matter, in all its vulnerability, defines sound art's path of ongoing investigation. Her physical waning serves as a touchstone of the risks involved in privileging resonance over instrumental reason and strategic self-management.

Echo's fate was to take on the sounds and vibrations of others, to take them in, resonate with and resound them, without being able to really reciprocate what arrived at her existential doorsteps. Echo's unwanted charity, her curse, is sound art's laboratory of sonic experimentation, its challenge, a haunting memory against which it defines its ambitions and affordances. Sound art, in all its heterogeneity, asks its audiences—like Echo—to take on heard and unheard sounds, to resonate with their environments, without establishing contracts or expectations of reciprocity. What defines the project of sound art is the effort to explore the potential of sound and vibrations to give place to something strange, unknown, and unexpected without anticipating any possible payback; to put the borders of subjectivity, of human agency and interaction, of the familiar and hegemonic, to the test in the absence of exactly knowing what the arrival of unheard (of) sounds will bring. Therefore, when we approach works such as *The Visitors* as sound art, we approach them as sites to probe the wonders and perils of hospitality—to probe sound's capacity and limits to enable hospitable attachments (Figure 1.5). What, we ask, does it take to let audible or inaudible sounds come in and visit? How do we welcome the audible and inaudible to attach themselves to our lives and us to them? How can the arrival of vibrational forces brush against both the singularization of bodies and the transformation of bodies into masses, against a seemingly seamless and deceptive integration of subject and object, subjectivity and world? And to what extent can and shall we respond to the arrival of unheard sounds as engines of retunement and metamorphosis?

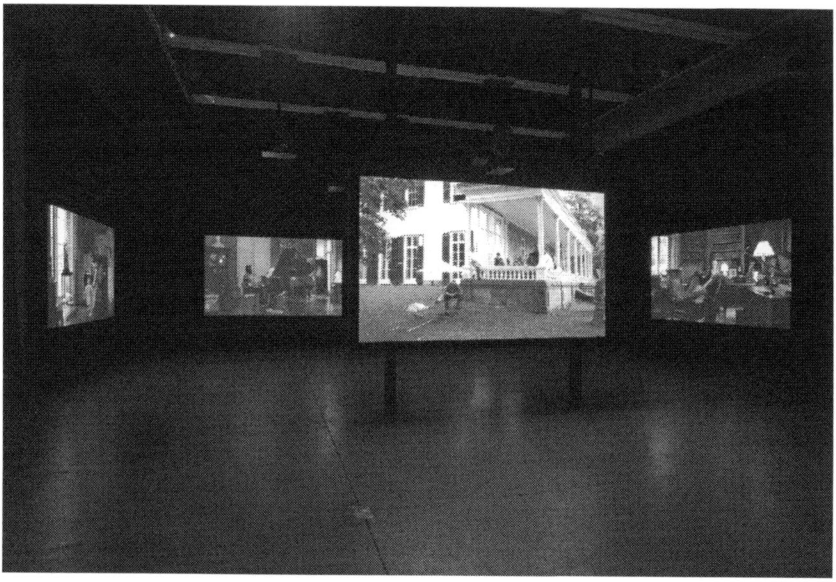

Figure 1.5 Ragnar Kjartansson: *The Visitors*, 2012. Installation view at Luhring Augustine, New York, 2013. See also Plate 1.

Struck by Juno's curse, Echo lost her voice and was doomed to live as a mere visitor amid her own life. It is time now to turn our attention to contemporary work that features sound's capacity to serve as a site and training ground of hospitality—work that explores resonance as a medium to overturn Echo's fate. The vexing calling of sound art is to inquire into the places, possibilities, and limits of resonant attachments amid inhospitable times. As it cuts across existing disciplines, labels, genres, and categories, sound art invites audiences to welcome the arrival of strange sounds and vibrations—the presence of visitors—as opportunities for emergence and transformation rather than as a threat and source of fear. Resonance is sound art's pivotal subject and object, its infrastructure of possibility and potentiality. It is as critical to the questions sound in art may raise as, say, the play of light to a photographer or the trace and promise of touch to a work of sculpture.

2

The Visitors

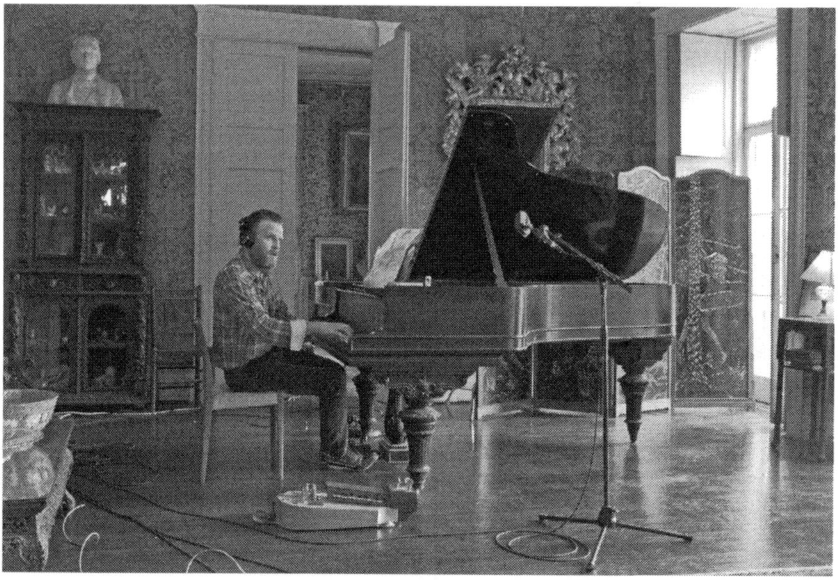

Figure 2.1 Ragnar Kjartansson: *The Visitors*, 2012 (Davíð Þór Jónsson). See also Plate 9.

A Pink Rose in the Glittery Frost

Nine screens. All except one display in extended long takes eight individual musicians at work, none of them at first sight sharing the same room with another, each of them wearing earphones, all of them feeding their sounds into nearby microphones so as to be heard by their fellow players. Though the cameras recording these musicians remain static, the images they produce are rich in detail and summon visual scrutiny. There is lots to look at and much to marvel about. The patina of old-word charm and former new-world affluence. The display of family heirloom and patronaged art. Convoluted gold frames, statues and busts, used and unused porcelain, eclectic furniture, floral wallpaper, shelves loaded with books no one has touched in decades, well-treaded hardwood floors. Seasoned wall paint, peeling almost everywhere and yet entertaining the eye with a wide spectrum of warm, feel-good colors. Unless you count yourself a staunch architectural modernist, each screen offers candy to the eye. Their luster draws us in, invites us to settle our gaze on comforts of old and consume our own nostalgia. They entice us to become sleuths perusing the surface of each image as if it were a treasure chest, a display case of unspoken mysteries and long-forgotten histories.

And yet, whatever we see in front of our eyes and at the edges of the visual field is soon colored, framed, and engulfed by probing sounds traveling across the installation space, and before we know it we'll linger at least as much with what we hear—a minimalist song that defies typical formats—as with the visual choreography of each and every of the installation's nine screens. Sixty-four minutes in length, this song swells to a few moments of glorious polyphony amid long stretches of repetitive languor. Its rather sparse lyrics are enigmatic, its sluggish pace refutes the visual business of each screen, and the instruments used to orchestrate the singing are as conventional as they at times assume the wondrous. We hear drums, pianos, cellos, guitars, and accordions. But we also witness the sounds of cheap acoustical guitars strummed while being partly submerged in a bathtub or of a pianist directly playing the strings of his instrument in true avant-garde nature. And, of course, there is the intermittent blast of an old cannon, visible on the installation's ninth screen together with a group of people gathered on the porch of an old mansion.

Unforgivingly slow, the song has little to stop listeners in different contexts from pressing their players' fast forward buttons. While some viewers might find it challenging to stay with the tune, others might find it equally challenging to take in the performance as a whole. The installation's nine screens are typically hung such that visitors cannot but fail to find one stable point of view allowing them to take in everything at once. In order to follow the often minimal gestures by which individual players accent the song and drive things forward, listeners will need to constantly adjust their perception and reposition themselves in the installation space. This need to wander around is further complicated by the fact that each screen's speaker only transmits the sounds of the musician visible in the image and thereby, to some degree, screens out the sound of the others; and by the fact that all but one of the nine screens will turn blank toward the end of the performance.

Music bordering the boring, images evading comprehensive visibility, audiovisions that encourage mobile and rather restless audience positions: the ambience of Ragnar Kjartansson's *The Visitors* seems to offer a perfect playground for contemporary art viewers who meander—like channel zappers, window shoppers, or digital multitaskers—from one attraction to another, yet rarely commit their entire attention to the works on display, let alone try to meet these works' durational extensions. Since its premiere at the Migros Museum in Zurich in 2012 and its first showing in the United States at New York's Luhring Augustine Gallery in early 2013, the work has been presented to considerable fanfare in museums around the world as prestigious as the Thyssen-Bornemisza Art Contemporary in Vienna, the Broad in Los Angeles, the Guggenheim in New York City, the Institute of Contemporary Art in Boston, the Turner House Gallery in Penarth, the Hirshhorn Museum in Washington, DC, the Frist Art Museum in Nashville, and the San Francisco Museum of Modern Art. Yet even most casual research on social media sites will quickly reveal that audiences have greeted *The Visitors* not with the twitchy agitation and fragmented edginess of 24/7 screen culture, but with a curious sense of mesmerized immersion, allowing the piece to wash over and away the unquiet of the everyday. *The Visitors* may cater to viewers on the move, but it nevertheless manages to move its visitors in remarkable ways, to draw them into its space of performance, to be hospitable to viewers with the most diverse backgrounds, preferences, and expectations. Over and over again, visitors report watching the piece more than once in a row, feeling arrested by its hypnotic pace and emotional registers, responding with both lasting melancholia and joy. Time and again, audience members confidently eschew the language of critical detachment and applaud the piece's spellbinding nature, its capacity to suspend the present's pressures of disruption and inattention, its unexpected ability to tear them away from the distractions of the everyday and literally move them to tears (Figure 2.2).

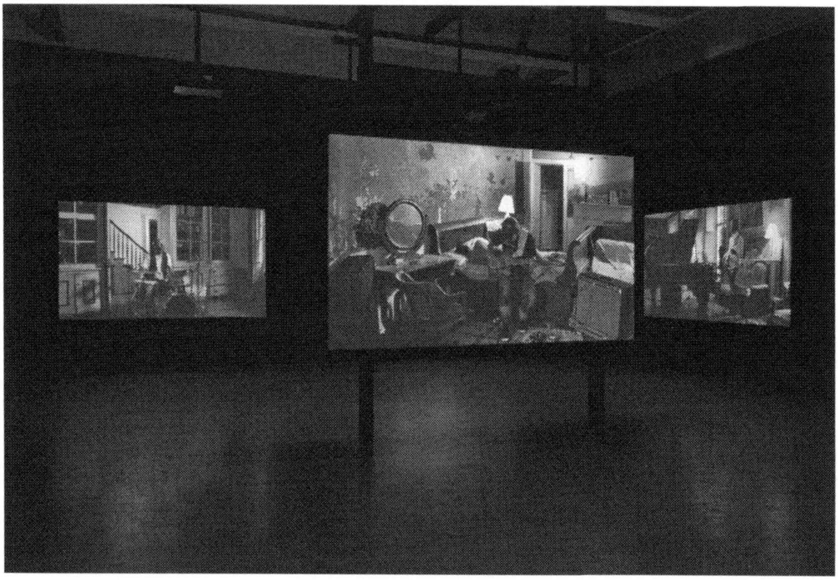

Figure 2.2 Ragnar Kjartansson: *The Visitors*, 2012. Installation view at Luhring Augustine, New York, 2013.

The Visitors opened in New York during a period that caused critics such as Seth Kim-Cohen, as they faced a series of exhibitions by artists as diverse as James Turrell, Brian Eno, Doug Atkin, Robert Irwin, Janet Cardiff, and Susan Philipzs, to rally against what they perceived as a precarious rise of the ambient: art approaching the aesthetic through sound's presumed ontologies and epistemologies, aesthetic practices that privileged perception over critical engagement and thereby discarded the legacies of forty or so years of conceptualism, discourse, and criticality. The ambient, Kim-Cohen argued, "in both its sonic and visual incarnations, describes a closed system. The sites of transmission and reception are identical. Nothing changes, nothing moves. It is ascetic and abstinent. Its apparent (desired) purity is but an abnegation of participation in the social, communicative, and critical realms."[1] Though *The Visitors* was not on Kim-Cohen's critical radar, it might have fit his bill as well. After all, Kjartansson's choreographed flow of sounds creates a quasi-auratic ecology of listening which, as confirmed by its visitors, seems to prioritize affect over vigilant criticality, resonance over reason, immersive enchantment over participatory engagement, consumption of what "simply is" over artistic

[1] Seth Kim-Cohen, *Against Ambience and Other Essays* (New York: Bloomsbury, 2016), 29–30.

practices that make us active, informed, and hyperaware of the power dynamics of the present. Seen as analogues to the prominence of object-oriented ontologies and speculative realisms in the realms of theory, the affects stirred by *The Visitors* around 2013, in the perspective of critics such as Kim-Cohen, qualified to be considered as effects of ambient shallowness at best, as politically and aesthetically regressive kitsch at worst.

There is no doubt something ascetic about the design of *The Visitors*. The song Kjartansson and his friends intone has no clear trajectory, the images manically and without any single cut focus on one performer each, the cameras' frames in eight of the projections never move an inch, the installation as a whole aspires to deliver no more than a record of a singular act of collaboration, a sculpture of sound in space. But does nothing change and move here indeed? Does the piece, with its deliberate distribution of sounds and images, its construction of audible space and ambient audition, stamp its production directly onto the perception of its audience? Or does *The Visitors*' ecology of sound in all its asceticism aspire to what Kim-Cohen himself considers after all the attributes of relevant art, namely to produce new and quite real relations "between the artist and the audience, between the artist and the materials, between one audience member and another, between each audience member and the collective audience, between each of these actors and institutions, between the present and history, between one artwork and another"?[2]

The underlying tenor of what visitors of *The Visitors* report when registering their response to the piece are experiences of resonance at both the somatic and the affectual level. They describe how the installation's spreading of sounds physically engages their bodies, as much as they communicate the way in which the mantra-like aspects of the tune produce powerful emotional echoes, ecstatic states of amazement and empathy. Hallucinatory and self-abnegating might be one set of words to conceptualize such effects, while wondrous and reverberating with unguarded hospitality might be another. It is easy to dismiss *The Visitors*' resonant vibrations as sentimental kitsch from the vistas of academic criticality. It is a much harder task, however, to explore exactly what happens when music moves people on the move, when gallery-based ecologies of sound manage to tear people out of the everyday and nevertheless touch upon their lives. No effort to understand the choreography of resonance in *The Visitors* can therefore do without inquiring into the nature, meaning, and historical index

[2] Kim-Cohen, *Against Ambience*, 13.

of an aesthetic tear so powerful that it can produce real tears: be they tears of sadness or of joy, of loneliness or of being together, of hospitality or receptivity, be they tears liquifying the hardened boundaries between subjects haunted by pressures of self-optimization and objects we have come to consider as numb and entirely pliable matter, or be they tears that, like a pink rose in the frost, express the seemingly impossible, namely a simultaneity of multiple and at first contradictory affects and emotions, in this case a messy blend of elation and melancholy, separation and sociability, interiority and exteriority.

A Diamond Heart and the Orange Red Fire

The Visitors was recorded in August 2012 at Rokeby Farm in upstate New York, halfway between Rhinebeck and the campus of Bard College along the Hudson River. Registered as a historic landmark that offers spectacular views of the Hudson Valley and the Catskills, the farm's mansion was built in 1815 on land granted to Colonel Pieter Schuyler in the 1680s. Though it continues to be inhabited by some of the heirs of the Astor and Livingston families, the 43-room estate today also serves as an event space and an artists' residency. Its rather weathered raggedness, radiating "old-world-ness" and memories of the Gilded Age, has come to inspire numerous artistic experiments and nonconformist extravaganzas, including the notorious staging of frequent pageants blending different spiritual traditions from Poland to the West Indies and from Africa to Tibet. As a 2010 profile in the *New York Times* put it: "Populated by a colorful but mostly impecunious cast of Livingston and Astor descendants—who are struggling, sometimes with each other, to keep the house from falling down while tending to their own deeply individual destinies—Rokeby is a study in contrasts, a lively dialogue, as one inhabitant put it, 'between the creatives and the historians.'"[3]

Known for his rhapsodic leaning toward melodramatic excess and the bohemian, Icelandic performance artist and musician Ragnar Kjartansson discovered Rokeby in 2007 when he posed for two days as a Hudson River School landscape artist, creating paintings and photographs of himself that were later exhibited at Bard's art museum. Upon his return in 2012, he brought seven musician friends along—six of them with Icelandic background, all of them with considerable repute across

[3] https://www.nytimes.com/2010/07/22/garden/22hudson.html.

various genres of music making—in order to intone *Feminine Ways*, a poem written by Kjartansson's ex-partner Ásdís Sif Gunnarsdóttir in response to their recent separation. After practicing the tune and its repetitive patterns for a few days, *The Visitors* was shot in one take by nine different cameras simultaneously, most of them situated in different rooms of Rokeby's appealingly run-down mansion. As much as it registered the artists' state during their prolonged stay at Rokeby Farm, the work's title also referenced the eighth and final album of the Swedish pop quartet ABBA, which had sealed the group's parting in 1981 while at the same time allowing ABBA's music to achieve a more complex and mature idiom. Some of the mansion's permanent inhabitants appear in Kjartansson's final video, as does the chef catering to Rokeby's far-travelled visitors' diverse tastes. We see Livingston heir and Rokeby owner Richard Aldrich fire an old cannon. We hear his wife, Polish artist Ania Aldrich, join what could be considered a chorus on the porch. And we witness the project's cook, Jacqueline Falcone, mostly sleeping in a bed behind guitarist Ólafur Jónsson, mimicking a famous seventh-century painting by Diego Velásquez, entitled *Rokeby Venus*, whose title has nothing to do with the farm's name but whose exhibition history has gained notoriety when the painting's surface was badly damaged in a suffragette attack in 1914 at the National Gallery in London.

The visual score developed and used to perform *Feminine Ways* at Rokeby Farm resembles the layout of a board game: *Chutes and Ladders* without vertical pathways and nasty traps, *Sorry!* without allowing players to progress in circles (Figure 2.3). It contains about thirty hand-drawn rectangular boxes, filled with minimal notations about dominant chords (dm, am, A, B, G, D) and abbreviated entries indicating certain moods and tempos. Written notes right below these boxes hint at leading instruments and identify the names of specific musicians meant to solo during this segment or simply take center stage. A number of drawings—stone bridges, a green train, an orange whirlpool (?), two explosive "Búmm"s—mark segments of retardation or transitional spaces, including Aldrich's firing of the cannon. Three inserted drawings of a radiating star with the words "D Stars" specify the intoning of a particular line of the poem about which I will have more to say later. With the exception of a few numbers indicating cycles of repetition, none of the boxes or drawn elements provide any clues about duration and intensity, thus leaving ample space for progressive adjustment and temporal (re)alignment. When compared with the actual performance, the score seems to elide the video's final minutes during which all performers—singing, walking, dancing—parade down Rokeby's lawn

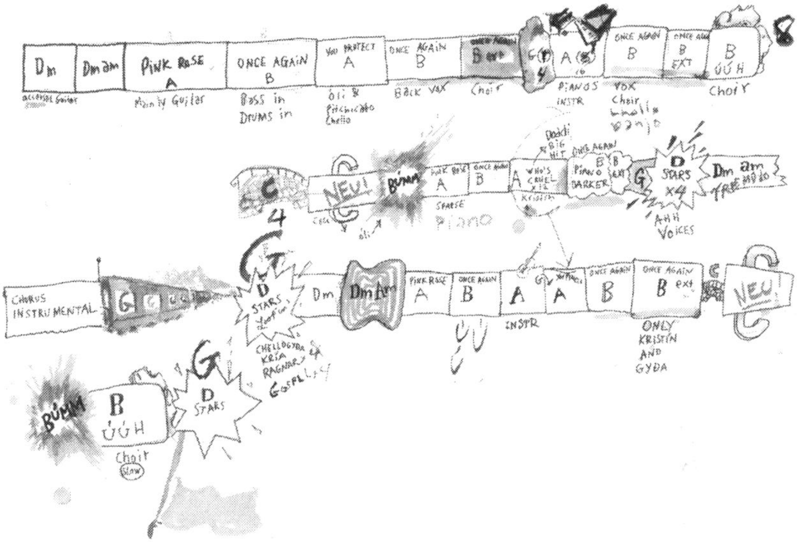

Figure 2.3 Ragnar Kjartansson: *The Visitors* sheet music, 2012.

toward the Hudson. Like the layout of a board game, the score offers a scaffold for numerous possible moves and engagements, a container at once opening and circumscribing future action. While the game's rules remain largely invisible, space for contingencies and interpretations abound. No subsequent activation of the plan will ever result in the same game, with the score's grid enabling infinite numbers of possible enactments and interactions. Gamers playing this game therefore cannot but carefully attend to the moves of other players in order to chime in at appropriate moments and help everyone to understand when and how to reach the next box, to enter the performance's next room.

Umberto Eco once considered graphic scores as "construction kits"[4] handing most of what matters about a musical piece to divergent actualizations of its performers. Extending Eco's thoughts on musical indeterminacy and the improvisational, Christoph Cox writes:

> Shattering the temporal logic of the traditional score, graphic scores are no longer virtual containers of sound that preserve, register, or record a sonic

[4] Umberto Eco, "Poetics If the Open Work," in *Audio Culture: Readings in Modern Music*, eds. Christoph Cox and Daniel Warner (New York: Bloomsbury, 2017), 243. For a fascinating collection and discussion of various graphic scores and notations, see Theresa Sauer, *Notations 21* (New York: Mark Batty Publisher, 2009).

> form or event that precedes them or prescribe a determinate performance that follows from them. Their orientation is entirely futural. Graphic scores are prompts or provocations for the production of new sonic events without a prior model.[5]

The Visitors' graphic score bears this out. Unlike the musical scores of Western art music, which presuppose conventional correspondences between visual symbols and acoustical notes, Kjartansson's *Chutes and Ladder* score defies clear and stable relationships between image and sound. It defines the visual and audible content of music as two separate channels and undermines traditional hierarchies between author, work, and music's actualization. Rather than authorize the performance, the score serves the role of a visitor hosted by the music and its unpredictable instantiations.

As much as Kjartansson and his friends when recording *The Visitors* at once enjoyed Rokeby Farm's hospitality and gave space to memories of ABBA's historical breakup, so does the playing of *Feminine Ways* express a rather unique gesture of offering a space to something having turned strange and other. Written after their separation, Gunnarsdóttir's poem does not simply register a chilling atmosphere of estrangement between former lovers, but also renounces possible hopes for reconciliation. It melancholically embraces a new state of withdrawal and solitude while protesting any effort of the erstwhile partner to project his cruelty onto the lyric I. In its final stanza, the poem in fact conjures a dramatic scene of destruction and death, a turning cold of what once provided light and warmth. The poem's last lines offer no prospect of reversal, no hope that any kind of action or intervention could turn around the course of time. All that's left after love's celestial burst is to let go, to abandon our stubborn belief in the fixability and non-finality of things human.

> *Feminine Ways*[6]
>
> A pink rose
> In the glittery frost
> A diamond heart
> And the orange red fire
>
> Once again I fall into
> My feminine ways

[5] Cox, *Sonic Flux*, 63.
[6] Reprinted with the kind permission of Ásdís Sif Gunnarsdóttir.

You protect the world from me
As if I'm the only one who's cruel
You've taken me
To the bitter end

Once again I fall into
My feminine ways

There are stars exploding
And there is nothing you can do

It is tempting to think that *The Visitors* violates the very instruction spelled out in the lyrics. Isn't to orchestrate and sing *Feminine Ways* to ignore the poem's final injunction to let go? Doesn't the whole performance seek, furtively and futilely, to inhale with life what has expired for good already? Doesn't the very act of singing the poem aspire to halt—and denounce—the poem's vision of *falling* and thereby reenact what may have caused the tearing away from one another in the first place?

Kjartansson himself considered the effort of giving place to the voice of his former partner as an exercise of feminine nihilism.[7] For him, to channel her distress, her anger about his cruelty, her plea to accept the separation, in fact her finding new comfort in dissolving—falling away from—the hardened subjectivities of former conflicts, articulated his own act of withdrawal, of recognizing, by hosting it, what has become and desires to be and remain strange. To sing *her* poem, in his view, was to acknowledge her claim to fall away, her effort to draw boundaries, her renunciation of a joint future. It was to mark the beginning of a future in which no joint future existed anymore. As a result, we may thus see the performance as an act of aesthetic ventriloquism, a calling forth and play with different voices; as a form of homeopathic cure, swallowing what causes pain, not in order to master it, but in order to learn how to move with it, to ease its burden by moving and thereby resituating it; as an enactment of (male) mourning in Freud's sense, a conscious dealing with grief to prevent painful losses and pathological forms of melancholia.

Any of these strategies no doubt runs the risk of being seen as a wicked strategy of incorporation and cooptation, of amplifying rather than recalibrating former friction, of silencing the other's voice in the very act of featuring it. It is indeed tempting to approach Kjartansson's act of ventriloquizing the voice

[7] https://www.luhringaugustine.com/news/ragnar-kjartanssons-the-visitors.

of a former lover as a modern replay of Echo's mythological marginalization; and Kjartansson's position in the bathtub as a surreptitious reenactment of Narcissus's unique relationship to water. Instead of trying to dissect the mess of an artist couple's breakup, however, I instead want to approach *The Visitors* as a paradigmatic model of how to think in more general terms about the dynamics of both hospitality and resonance in contemporary sound art, its efforts to treat sound as sculpture and emancipate music from traditional regimes of pitch, tempo, rhythm, timbre, and melody. Giving place to and resounding the (absent) voice of Gunnarsdóttir, the performers of *The Visitors* here inhabit a place in which they embrace their status as welcome guests at Rokeby Farm to probe resonant interactions across the mansion's separate rooms. Their performance of *Feminine Ways* investigates the work and mediation necessary to host, be touched by, listen to, and resonate with music's vibrations. And in so doing, their piece asks tough questions about what it means to be active or passive, agent or receiver, to speak or to being spoken, to fall or to catch, to be diamond or fire in the first place.

Once Again I Fall into My Feminine Ways

Similar to how we have come to misconstrue silent cinema as a cinema without sounds, so the modernist ideology of the museum as a white cube imagined gallery spaces as entirely muted, clear of any auditory distractions and ambient noises in order to train the viewer's attention exclusively onto the display of individual work.[8] Yet even before sound-based work entered contemporary gallery spaces, sounds of all sorts already populated and permeated art museums and disturbed the white cube's ideological investments into ideas of purity. The influx of video, installation, and sound art since the 1990s has added yet another level to this, whether it is embraced as a welcome diversification of exhibitable objects or loathed as an assault on desired economies of attention. "Upon entering almost any contemporary gallery space," Caleb Kelly mused in 2017, "we hear sounds emanating from TV monitors, projection spaces, computers and in headphones, alongside the daily sounds made by gallery staff, art patrons, the gallery bookshop and so on. The gallery is not the hushed space it was once

[8] Brian O'Doherty, *Inside the White Cube: The Ideology of the Gallery Space*, expanded edition (Berkeley: University of California Press, 2000); see also Hans-Ulrich Obrist, *A Brief History of Curating* (Dijon: Les presses du reel, 2008).

imagined to be, but filled with noisy, quiet, disruptive, overlapping, discrepant, loud, brutal, pretty, aggressive and/or harmonious sound."[9] Some of this is often hard to take in, not least of all because the design of most museums manages sound containment badly and provides little to prevent individual sounds from interfering with the reception of other artwork in adjoining galleries. One need not be a champion of the white cube and its imaginary in order to feel bothered at times by the cacophonies of contemporary museum spaces.[10]

What is important for our context here, however, is to recognize two aspects. First, what we understand as sound art, as "sound (,) art," activates wanted and unwanted aspects of sound to shape, inflect, expand, or complicate the very spaces in which art is exhibited. It generates deliberate or plays with unexpected echoes, not simply in order to occupy a given place as if it was a neutral container, but to produce space as an environment of aesthetic experience. Sound in video installations such as *The Visitors* thus actively disturbs the very vision that informed the spatial politics of white cube ideology. It understands, or makes us understand, space as both a temporal and relational category of experience—as, in Doreen Massey's words, a "dynamic simultaneity, constantly disconnected by new arrivals, constantly waiting to be determined (and therefore always underdetermined) by the construction of new relations. It is always being made and always, therefore, in a sense, unfinished (except that finishing is not on the agenda)."[11] Second, in upsetting the white cube's hushness and idealized picture of siloed subjectivity, sound in works such as *The Visitors* tends to emphasize communal or collective modes of reception over visions of individualized attention, criticality, or absorption. It may sweep audiences off their feet, but its echoes or reverberations, its discordant noises or harmonious rhythms, often aspire no less than to rearticulate the relational dynamic of its recipients among each other. Noisy or not, works such as *The Visitors* engage the mode of sound art when their acoustical dimensions seek to reengineer affects that may bind people, restructure their audience's relation to space and time, and

9 Caleb Kelly, *Gallery Sound* (New York: Bloomsbury, 2017), Kindle file.

10 For more on the acoustical turn and challenges of contemporary exhibition spaces, see, among many others, Peter Gruneisen and Bob Hodas, eds., *Soundspace: Architecture for Sound and Vision* (Basel: Birkhäuser, 2003); Steven Connor, "Ears Have Walls: On Hearing Art," Lecture given in the series *Bodily Knowledges: Challenging Ocularcentricity*, Tate Modern, February 21, 2003, http://www.stevenconnor.com/earshavewalls/; M. Zisiou, "Towards a Theory of Museological Soundscape Design: Museology as a Listening Path," *Soundscape*, 11.1 (2011): 36–8; Rogers, *Sounding the Gallery*; Nikos Bubaris, "Sound in Museums—Museums in Sound," *Museum Management and Curatorship* 29.4 (2014): 391–402; and various contributions to the special issue on the "Sonic," in *Curator: The Museum Journal*, 62.3 (July 2019), edited by Kathleen Wiens and Eric de Visscher.

11 Doreen Massey, *For Space* (London: Sage, 2005), 107.

in this way explore what it might mean to be a subject in the first place. Sound art's echoes remind us of the fact that neither space nor time, perception nor subjectivity ever come in the singular. It pushes against normative models that want to shelter the willfulness of singular agents from unwanted interferences, distractions, and disruptions.

Jean-Luc Nancy once argued that we cannot uncouple sound from its echoes: even if you sit in an anechoic room, sonority is largely unthinkable without the immanence of resonance:

> Reflection requires a reflective surface, in principle external to the visible thing. Resonance is inside of sound itself: a sound is its own echo chamber, just as it is its own timbre, its overtones, and what is called its color. The ancient Greek *êkhéô*, from which "echo" comes, means "to make noise" as well as "to re-sound": it signifies precisely "to return a sound." A sound is always "returned," restored: it is restored from itself to itself. A sonorous body that is struck *returns* the blow by the sound that is the vibration of the blow itself. Sound is at the same time *struck* (pinched, rubbed, breathed, etc.), *returned*, and *heard* (entendu, understood) in the precise sense that *it is understood* (s'entend) or that it *makes itself heard* (se fait entendre): and for that, in that, it *listens to itself* (s'écoute).[12]

Though the spatial setup has been somewhat different at each exhibition site, *The Visitors* is designed to at once enable and probe what Nancy calls the return of sound and the immanence of resonant matter. In most installation settings, seven of the nine screens will be lined up along the gallery's four walls, in pairs of two and with one mounted by itself. The remaining two screens take up some space in the middle of the gallery, installed back to back. Each screen is connected to a loudspeaker broadcasting sound solely associated with the performance of the musician visible on the screen in question. What you hear when approaching any of the nine screens, in other words, is the sound that was exclusively captured by the microphone you see in the respective image and was during the recording session transmitted to the earphones of all other performers. The 64-minute loop will commence with one image after the next being switched on by a crew member as each of the performers readies themselves for the recording. It will end with all musicians gathering in one of the rooms while each of the other screens will become blank and silent again and the camera

[12] Jean-Luc Nancy, "Foreword: Ascoltondo," in *Listen: A History of Our Ears*, by Peter Szendy (New York: Fordham University Press, 2008), Kindle file.

recording the remaining image will eventually leave its static position so as to capture the group's procession down the lawn toward the Hudson.

These details matter for a number of reasons, all of which are important to understand how *The Visitors* offers a unique echo chamber in which sounds can listen to themselves and audiences are invited to stitch together sounds and images in space in ever-different ways. In her seminal book on the aesthetics of installation art, philosopher Juliane Rebentisch argues that contemporary art's appeal to mobile, unpredictable, and often highly contingent acts of viewership calls for a fundamental revision of modernist concepts of autonomous art.[13] Rather than abandon the idea of the specificity of the aesthetic altogether, however, Rebentisch embraces installation art as a unique laboratory to develop new understandings of art's ethical and political dimensions that place the relative autonomy of aesthetic experience at the core of what it takes to negotiate the contours and fabrics of the social today. *The Visitors* echoes Rebentisch's considerations, offering an experimental site of restitching the fabric of social relations. First, unable to ever take in all screens at once even when all of them have something to show, viewers of *The Visitors* find themselves void of any stable, ideal, and sovereign viewing position. In spite of the music's sweeping and absorptive qualities, the piece's spatial choreography encourages viewers to explore the installation's set up as roaming subjects adding individual and fragmented vantage points on the go without ever affording some sense of totality and synthetic integration. Second, though offering a quasi-Platonic cave of visual and acoustical projections, *The Visitors*, by engaging viewers to reposition themselves in the gallery space, envisions its visitors as embodied viewers whose seeing and listening is restricted neither to eyes nor ears but involves tactile and kinesthetic aspects as well. People here listen with their legs as much as they see with their feet. Third, the experience of watching *The Visitors* asks the viewer to perform a certain kind of labor: the work of coordinating one's own often mobile viewing with the mobility of other spectators co-present in the gallery space. No viewer here is ever in full control over his or her own act of viewing. Moreover,

[13] Juliane Rebentisch, *Aesthetics of Installation Art* (Berlin: Sternberg Press, 2012). For more on the aesthetics of contemporary installation art, see Claire Bishop, *Installation Art: A Critical History* (London: Tate, 2005) and *Artificial Hells: Participatory Art and the Politics of Spectatorship* (London: Verso, 2012); and on shifting concepts of viewership in multichannel exhibitions spaces, Kate Mondloch, *Screens: Viewing Media Installation Art* (Minneapolis: University of Minnesota Press, 2010); Gabriele Pedulla, *In Broad Daylight: Movies and Spectators after Cinema* (London: Verso, 2012); Erika Balsom, *Exhibiting Cinema in Contemporary Art* (Amsterdam: Amsterdam University Press, 2013); and Andrew V. Uroskie, *Between the Black Box and the White Cube: Expanded Cinema and Postwar Art* (Chicago: University of Chicago Press, 2014).

because viewers, due to the demand to resituate themselves, tend to possibly block each other's view, the audience experiences a profound need to reckon with and respond to the viewing of other onlookers so as to create an operative equilibrium, a resonant network, of collective perception. Bodies shift so as to improve sightlines, but also may forgo perfect viewing angles for individual or entire groups of screens so as to allow adjacent spectators to see something, too.

Fourth, and foremost however, with its nine separate and distributed channels of sound projection, *The Visitors* cannot but position the listener as an active sampler of auditory input. In this, the experience of visiting *The Visitors* is reminiscent of the way in which the forty speakers of Janet Cardiff and George Bures Miller's installation *Forty Part Motet* (2001), a reworking of Thomas Tallis's 1573 *Spem in alium*, have the work's listeners actively explore different auditory points of view, their motion from one speaker or one group of speakers to another thus emerging as a way of constructing the music in the first place. Similar to how in Cardiff/Miller the listeners' walking animates the music and constantly reshuffles the sculptural relationships between Tallis's polyphonic voices,[14] so do *The Visitors*' nine speakers encourage listeners and viewers to become peripatetic examiners, using their very bodies as media to probe ever-new arrangements of auditory and visual input, ever-different ways of how to resonate with and fall into *Feminine Ways*. To roam in, amid, through, and across *The Visitors*, its sounds and sights, is to embrace walking as a technique to become a DJ of sorts, reassembling the original performance—its distribution of sensible material—with each step of our bodies and amid a crowd of other onlookers and listeners.

It is questionable whether it is really possible to listen to our own listening.[15] And yet, *The Visitors* offers a space in which listeners can kinesthetically explore nothing less than the plasticity of audition—the way in which the echoes and resonances of sound can be arranged through deliberate and contingent acts of hearing. While to listen to our own listening might in fact mean to hear nothing at all, when entering the space of *The Visitors* we certainly experience ourselves as arranging and rearranging the piece's acoustical and visual relations and thereby observe the work it takes to establish resonant relationships between listener and listened. To examine the nexus of movement and being moved is thus as much part of the ecology of listening to *The Visitors* as it is to experience listening as something strangely straddling the lines between the singular and

[14] Lutz Koepnick, *On Slowness: Toward an Aesthetic of the Contemporary* (New York: Columbia University Press, 2014), 239–44.

[15] Peter Szendy, *Listen: A History of Our Ears* (New York: Fordham University Press, 2008), 142.

the communal, the private and the social. Like the players involved in jointly intoning *Feminine Ways* at Rokeby Farm, for us to enter the space of *The Visitors* is to probe what it takes and means to attend to—stretch oneself or being stretched out toward—auditory objects, and to witness how our listening in fact co-constitutes these objects in the first place. It is to find oneself amid conditions in which the experience of listening unsettles preconceived notions of the merely singular and entirely communal, the active and the passive, recipient and work. Neither do we listen here with and as one single body nor do we encounter the space of our listening as stable and unified, as something whose elements we could, metaphorically speaking, stick in our pocket and, like bad guests, simply take home. To attend to *The Visitors* is to recognize that mere additions of individual perspectives will not neatly result in anything total and conclusive; that no listening exists without objects holding a listener's attention, but also without other listeners whose listening inflects our own; and that orthodox templates and often engendered juxtapositions of active and passive viewership, the former praised as a gateway to critical and self-reflexive participation, the latter dismissed as the signature of consumption-driven numbness, require serious reconsiderations, for reasons unfolded in much greater detail in later chapters of this book.

The spatial choreography of sounds and sights of *The Visitors* evoke a co-presence of contradictory stances and seemingly incommensurable approaches. We listen as much as we sense our listening; movement and being moved go hand in hand; intimacy and publicness cross paths; feelings of immersion run parallel to the need of attuning our steps to those of other audience members. The undulations of auditory material, argues Brandon LaBelle,

> do much to unfix delineations between the private and the public. Sound operates by forming links, groupings and connections that accentuate individual identity as a relational project. The flows of surrounding sonority can be heard to weave an individual into a larger social fabric, filling relations with local sound, sonic culture, auditory memories, and the noises that move between, contributing to the making of shared spaces.[16]

LaBelle's stress on the relational dynamics of sound urges us to think through why visitors of *The Visitors* frequently report the feeling of being swept off their feet by the undulations of *Feminine Ways*, affect seemingly submerging

[16] Brandon LaBelle, *Acoustic Territories: Sound Culture and Everyday Life* (New York: Continuum, 2010), xxi.

any sense of self, the song's sonority pulling them into the music's flow. LaBelle alerts us to the fact that more is at stake than simply thinking of their listening—our falling into intimate absorption amid a semi-public community of other listeners—as an instance of mindless submission, of therapeutic narcissism or of sentimental kitsch. For now let me simply say this: similar to how Kjartansson and his friends themselves tuned and retuned the relational logic of shared space at Rokeby Farm, so does *The Visitors* grant its visitors a space to examine the extent to which resonant modes of subjectivity *and* communality can suspend contemporary mandates of 24/7 self-management and strategic interaction. The installation's particular ecology of sounds and sights invites its audience—in form of an aesthetic promise—to engage in alternate forms of sociability in whose context experiences of awe, wonder, and enchantment need not fear to underwrite anti-progressive politics or populist desires for social enclosure.

You Protect the World from Me as if I'm the Only One Who's Cruel

It has become a commonplace in contemporary sound studies to emphasize, as I just did, the participatory dimension of listening and thereby answer in the affirmative George Berkeley's famous eighteenth-century question whether trees falling in a forest need a listener to make a sound. In the wake of in particular Christopher Small's studies on musicking,[17] no one will also wake any sleeping dogs anymore when insisting that the performative contexts, institutional and architectural settings, culturally specific practices and expectations, and the entire apparatus of production, recording, storage, distribution, and dissemination may matter as much as the sound itself to understand what music does to its listeners and how it shapes various kinds of relationships. But it would nevertheless be silly to protect the reader from learning more about, on the one hand, the formal elements of the music itself and how even some of its most self-absorbed sonic features may serve as sundials of much larger dynamics, and about, on the other hand, the composers, players, and performers largely responsible for delivering vibrational objects to our senses. So, while we deal with the performance's musical features at various later junctures of this book,

[17] Christopher Small, *Musicking: The Meanings of Performing and Listening* (Middletown, CT: Wesleyan University Press, 1998).

here's a brief overview of Kjartansson's seven collaborators, followed by a brief reflection on the art and ideology of collaboration in an age of networked labor.

Davíð Þór Jónsson (Figure 2.1 | Plate 9) serves as one of the two pianists. Like everyone else except Shahzad Ismaily a native of Iceland, Jónsson is a frequent collaborator of Kjartansson. In *The Visitors*, we see him at times playing the piano in unconventional ways, smoking fat cigars, contributing choral texture, and—perhaps more importantly—guiding all musicians with extended piano bridges from one module of the performance to the next.

Working mostly as an architect, *Ólafur Jónsson* (Figure 3.1 | Plate 6) plays the electric guitar. Similar to other members of the band, he has collaborated frequently with Kjartansson on earlier projects, after and parallel to extensive work with various local bands in Reykjavik. Together and behind him on the bed, we see Jacqueline Falcone as she sleeps through most of the performance and only gets up when Jónsson is ready to join the other band members in the piano room and leads her in stately fashion out of the frame.

Gyða Valtýsdóttir (Figure 5.1 | Plate 4) started her career as a member of Múm before moving to Switzerland to study cello, eager to master both classical and experimental uses of the instrument. A collaborator with artists as diverse as Damien Rice, The Kronos Quartet, and Ben Folds, her internationally recognized recording *Epicycle* (2017) featured not only a constellation of pieces by Schubert, Schumann, and Messiaen but also more experimental composers such as Harry Partch and George Crumb.

Kristin Anna Valtýsdóttir (Figure 6.1 | Plate 5) is heard and seen playing the accordion and toward the end of the performance an acoustical guitar. Like her twin sister Gyða, she started out her career as a member of the Reykjavik electronic band Múm, moved on to play for Animal Collective and other notable New York-based bands, and has made a name for herself for solo shows blending shamanistic extravaganza and avant-garde experimentation. I will discuss the vibrancy of her high-pitched voice, as put to work in various solo, duet, and choral sections of *The Visitors*, later in this book.

Shahzad Ismaily (Figure 7.1 | Plate 7) is mostly playing the bass guitar in *The Visitors* but also some banjo. Born to Pakistani immigrants, he is based in New York. His background is both in jazz and rock; he has done compositional work for some film soundtracks, theater performances, and musical productions as well.

Located in one of the mansion's dining rooms, a staircase ascending to the house's second floor behind his drum set, *Þorvaldur Gröndal* (Figure 8.1 | Plate 10) acts as the group's percussionist. He is a trained ethnologist, with a special interest in rural Icelandic furniture design, yet has also served as a long-time collaborator of Kjartansson on numerous projects over the last decade or so.

Though it will become clear only more than halfway through the performance, *Kjartan Sveinsson* (Figure 9.1 | Plate 8) plays his grand piano and, less frequently, his electric guitar in the same room and in sight of Davíð Þór Jónsson. Sveinsson is known both as an art music composer—the ethereal *Der Klang der Offenbarung des Göttlichen* (2016) stands out amid his recordings—and as a former member of the Icelandic post-rock band Sigur Rós. His stylistic signatures lean toward the epic and sublime.

Last but not least, the porch (Figure 1.1 | Plate 2) gathers a number of crew members and Rokeby residents, including Richard and Ania Aldrich. They are mostly shown as listeners but also offer some choral background. Richard Aldrich's role, in addition, is to operate the mansion's cannon, presumably once located in the Forbidden City, yet later transported to the United States after the Boxer uprising in China around 1900.

The collaborative spirit of *The Visitors* should make us pause for a moment to think about the status of collaboration in our contemporary moment. The neoliberal economies of our present, while relying on the labor of self-managing singularities, highly value collaborative efforts of various kinds: the coming together of different human resources with different skills/ideas/visions/experiences to solve problems whose complexity beats the capacities of isolated individuals.[18] The idea of collaboration is often wedded to the fetish of the new and the entrepreneurial aura of disruptive innovation. It cherishes the ways in which collaborative efforts, whether they produce physical or digital objects, new forms of knowledge or spectacular experiences, break with habitual patterns of the past and set unique standards for the future. Collaboration is to the creative industries of the twenty-first century what Adam Smith's invisible hand was to the burgeoning capitalists of the seventeenth and eighteenth centuries. It registers the real as much as it masks it. It produces the very isolation of goal-oriented individuals it pretends to overcome.

What we witness in *The Visitors* is of a different order. The performance, at heart, features what I want to call resonant collaboration—a mode of human interaction fundamentally at odds with the neoliberal worship of creative innovation through coordinated self-interest. As profiled in *The Visitors*, resonant collaboration entails forms of cooperation in which the presumed passivity of listening turns out to

[18] Mark Terkessidis, *Kollaboration* (Frankfurt am Main: Suhrkamp, 2015).

be as important to the musical production of human relationships as any gesture of making sound itself. Rather than to fuse self-regard and goal-orientedness in the common chase for the next big thing, resonant collaboration here embodies modes of sociality in which durational structures of repetition, recursivity, flow, slowness, and circularity dispute our times' stress on disruptive innovation. *The Visitors* leaves little doubt that resonance requires labor. Collaboration is shown as a piece of work. But instead of feeding into a restless culture of fearful self-protection and industrious self-performance, the work of Kjartansson's band promotes the ability to care for and attune oneself to the lives of others as a vibrant medium of commonality, a medium that needs, takes, and claims time to upend the instrumental and cynical reason of the present.

You've Taken Me to The Bitter End

Let's press the rewind button for a moment. Karlheinz Stockhausen's 1964 *Mikrophonie I* (Figure 1.4) engaged two microphones as instruments not simply to amplify or inflect the rhythmic movements of two tam-tam players, but to produce sounds on their own and investigate the audible's acoustical unconscious. Similar to the work of a microscope, Stockhausen's two microphones were used by two performers to explore the sound of the manipulated tam-tams up close; capture specific pitches, timbres, and rhythms produced by implements attached to the percussion instruments (found objects made of glass, cardboard, metal, wood, rubber, and plastic); and record their own noises when being moved quickly through space following specifically scored gestures. The microphone's output was passed on to two additional performers who were seated in the middle of the audience and employed bandpass filters to feed the resulting sounds into the room's quadrophonic speaker setup. In its thirty-three different structural units—called "moments" in Stockhausen's process-oriented language of the time—*Mikrophonie I* was thus tasked to project normally inaudible vibrations and make them, in Stockhausen's own words, "audible by an active process of sound detection (comparable to the auscultation of a body by a physician); the microphone is used actively as a musical instrument, in contrast to its former passive function of reproducing sounds as faithfully as possible."[19] Similar to how in nineteenth-century medicine the stethoscope

[19] Karlheinz Stockhausen, "Mikrophonie I (1965), für Tamtam, 2 Mikrophone, 2 Filter und Regler," *Texte zur Musik* (Cologne: M. DuMont Schauberg), III, 57.

introduced new techniques of listening and precisely thereby produced new ways of understanding the human body,[20] Stockhausen employed his microphones to enable new modes of listening and unlock otherwise unheard vibrations for compositional processes. The medium—the techniques and technologies of mediation—became the message (Figure 2.4).

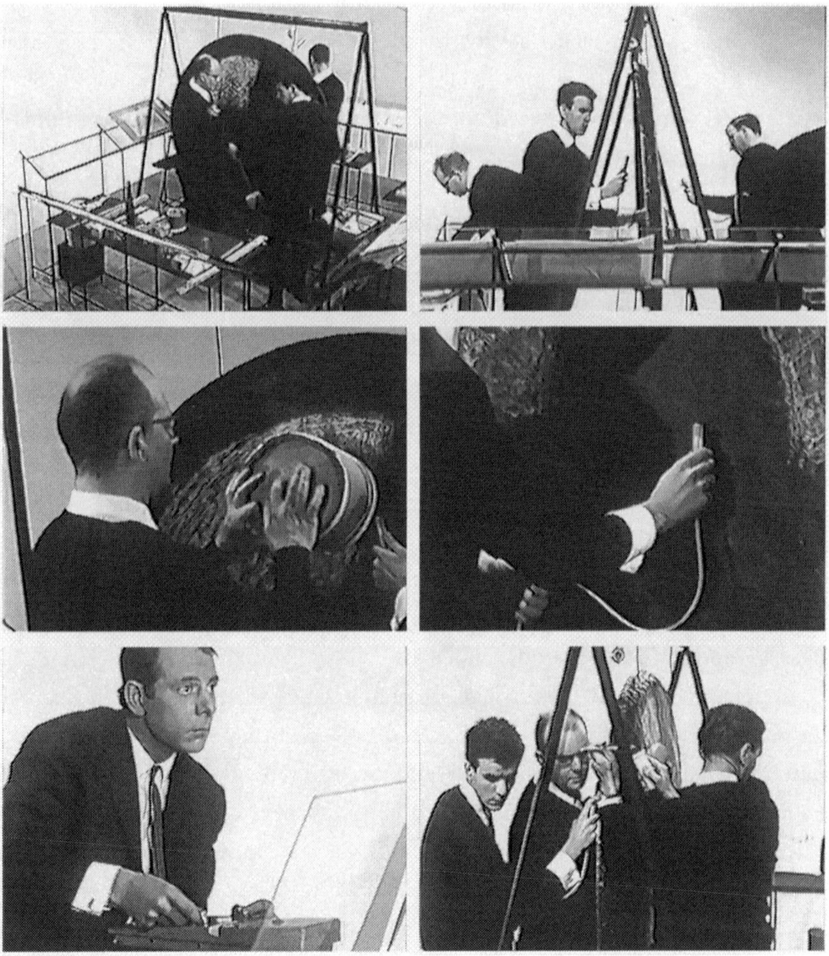

Figure 2.4 Karlheinz Stockhausen: *Mikrophonie I*, 1964.

[20] Jonathan Sterne, *The Audible Past: Cultural Origins of Sound Reproduction* (Durham: Duke University Press, 2003), 87–136.

In *The Visitors* no one moves, let alone hurls, swings, or whacks, any of the microphones visibly at work. They remain stationary, still and silent observers of what is performed in their vicinity, primarily relaying and echoing captured sound to other locations. Unlike Stockhausen's 1964 use of microphones as musical instruments and tools of creative investigation, Kjartansson's mics at first appear mostly as detached utensils whose technology is taken for granted and whose modalities of listening, therefore, require no further mentioning.

Nothing would be more mistaken, however, than to argue that *The Visitors* obscured the process of mediation to which the singing on screen and our listing in the gallery owe their very existence. Nothing of what we hear and see in *The Visitors* is shown as "simply" being there. None of what vibrates across the screens and resonates with our affects is shown in isolation from the elaborate process of production and transmission. And we are not invited to believe that the making, recording, transmission, and reproduction of sound here had no role in how *The Visitors* brings into play different relationships between collaborating artists, between artists and their audiences, between art and its institutions of display, between music and its materials, or between different audiences and members themselves. Kjartansson might sit in his bathtub like a modern-day Narcissus, as if to define the performance's apparatus of transmission as no different from how ancient myth presented the figure of Echo: as a merely passive receiver and recorder of sound, void of agency, a sole reflector of the audible. And yet, as staged in front of its nine cameras and microphones and projected onto its nine screens, *The Visitors* nevertheless is all about the intricacies of advanced mediation. It is about how audio technologies interconnect subjectivities across time and space, and about how microphones, earphones, and loudspeakers shape and are being shaped by specific approaches, agendas, skills, methods, and practices—in one word: techniques—of perception. Mediation is *The Visitors*' form and content—its crucial message, as seasoned McLuhanites would say. It does not matter how many tears its viewers shed and thereby cause critics to suspect the installation for privileging sentimentalist presence over critical reflexivity—*The Visitors*' key ambition is to investigate the possibilities, limits, pathways, and pleasures of resonant togetherness, enabled in spite *and* because of the very media that simultaneously connect and separate us.

There are probably as many definitions of the terms medium and media as there are professed media scholars. These definitions range from Communication Studies 101 concepts of media as neutral channels transmitting messages from senders to receivers, to John Durham Peters's neo-Heideggerian effort to consider media—reminiscent of how premodern cultures understood elemental forces such as earth, air, water, and fire—as "vessels and environments, containers of possibility that anchor our existence and make what we are doing possible."[21] This is not the place to embark on intricate discussions of the history and meaning of the concept of media, let alone parse the difference between using the term in the singular or its plural. What alone matters at this point is to insist that media, far from representing mere technological configurations, assemble material dimensions and human practices, technical features and cultural memories, norms, expectations, and utopias, hardware designs and institutional and social processes into single dynamics. Media belong to the realms of both nature and culture. They offer the material conditions—the quasi-natural, albeit historically constituted and inflected substrate—of the possibility of meaning-making, experience, and vibrant relationality.[22] Rather than embodying that which merely exists in-between, mediums and media provide infrastructures of human (and nonhuman) embodiment and sociality—that which enables us to become and be human in the first place.

This is just another way of saying that any discussion of media technologies, lest one want to subscribe to some form of media determinism, cannot do without distinguishing between concepts of technology and technique. Technique is what we do with given technologies, to live up to, to exceed, and to recalibrate their inherent protocols. In Alan Liu's words, "I define technique as a method or practice that goes beyond being an application to becoming a play on technology—as when we say there is play in the action of a machine part, not to mention in a musical instrument. Technique is both bound to and free from its technology. As such, it is generative of culture from nature."[23] Techniques take technologies to their sweet or bitter end. Or both at once. They bring them to play, play on and play with them, and in so doing transform the rules and practices technologies seem to prescribe. Techniques cause technologies to vibrate, at once infecting and being infected by what we typically consider a medium's specificity, its limitations, the logic of its setup. Techniques are to technologies what competitive divers are to diving platforms. The latter enable movements and performances—figures in space and time—whose skill and virtuosity will

[21] John Durham Peters, *The Marvelous Clouds: Toward a Philosophy of Elemental Media* (Chicago: University of Chicago Press, 2015), Kindle file.

[22] J. M. Bernstein, *Against Voluptuous Bodies: Late Modernism and the Meaning of Painting* (Stanford, CA: Stanford University Press, 2006).

[23] Alan Liu, *Friending the Past* (Chicago: University of Chicago Press, 2018), 11.

make attentive onlookers no longer think of the platform as merely a passive object, but as something endowed with a certain agency, receptivity, and life itself. Techniques spin technology into vibrant matter.

Kjartansson's *The Visitors* is a play on technology. The performers' peculiar technique of listening and singing playfully complicates two common views of contemporary media technology. It intermeshes what routinely is seen—and lamented—as radically opposed. Contemporary media offer platforms of 24/7 connectivity and interaction, in one view, so much so that we have lost the resources to endure solitariness, silence, and seclusion and instead, masochistically as it were, have come to experience our sleepless bondage to digital culture's economy of hyperattention, its protocols of ceaseless stimuli and interruption, as life's highest thrill. Twenty-first-century entertainment technologies create unprecedented forms of social atomization is the other view. They enclose the self in sensory and cognitive bubbles, undermine the conditions for meaningful discourse and encounters, and precisely thereby enable new forms of political division, populist manipulation, and totalitarian control.

The tear and tears, the sadness and joy, the affects and awe triggered by *The Visitors* registers the fact that neither of these two views really rings true on their own terms, not least of all because both eclipse the curious dialectic of technique and technology. At once bound to and freeing itself from the bonds of technology, *The Visitors* explores the extent to which the either/or of compulsive connectivity and speechless isolation need not be digital culture's final word; it puts to the test whether solitariness and togetherness can go hand in hand without producing pathological results. The performers' techniques of listening into each other at a distance, their exploratory gestures of establishing resonance and synchrony across spatial separations, our own techniques of roaming the installation, of tuning our steps and stops to its sights and sounds, and thereby of arranging ever-different versions of *Feminine Ways*—all this cannot but question the tenets of technological determinism, the fashionable, but ultimately utterly fatalistic view, that hardware trumps software and wetware.

Mediation, its pleasures and pains, are what makes *The Visitors* tick. To play on media technologies here is to recognize that we cannot do without them, not because media make us or we make media, but because what we have come to understand as the human is a medium itself: a resonance chamber of different technologies and techniques of humanization which at its best can afford to attune our desires for sociality and solitariness to each other.

There Are Stars Exploding and There Is Nothing You Can Do

Acoustical media in *The Visitors* allow performers and gallery visitors alike not simply to receive sounds from a distance but to lean into the sounds of others, to welcome sounds from afar to take up a place in the space of one's own listening. Far more than just connecting remote locations and subjects separated in space, the performance's microphones and earphones, as well as the installation's screens and loudspeakers, define the space of listening, in Jane Bennett's words, as assemblage: a site of "ad hoc groupings of diverse elements, of vibrant materials of all sorts . . . living, throbbing confederations that are able to function despite the persistent presence of energies that confound them from within."[24] Assemblages are not controlled by central heads of power. No single agent or material within their movements has sufficient power or authority to determine all the others. Some of their areas or nodes might be trafficked more heavily than others, yet none can claim a right or responsibility to monopolize the uneven distribution of power or influence across its surface. To lean into the sounds of others here means to forfeit one's own sense of mastery. It is to fall into something that exceeds one's control. It is to invite sounds to cross one's doorsteps and establish resonant relationships without knowing or erecting defensive walls against what this might do to the listener. Acoustical media in *The Visitors* redefine listening as the ability to touch upon and be touched by distant explosions of stars even if—or precisely because—there is nothing one can really do about their vibrations' impact, about the vital energy and force of events that, in actual fact, reach out to us from times past.

To lean and fall into something whose pulses might destroy us, to give place to the strange, unknown, or other without setting up defenses and expecting anything in return, to attend to distant catastrophes while renouncing any possible actions, interventions, or self-protective gestures: isn't all this, in political and moral as much as in psychological terms, at best deeply precarious, at worst utterly perverse?

Suspicions might be in order, but in this case they might run into the open door of a project whose very aesthetics of hospitality aspires no less than for us to rethink our very histories and habits of being suspicious, our failures to permit

[24] Jane Bennett, *Vibrant Matter: A Political Ecology of Things* (Durham: Duke University Press, 2010), 23–4.

certain encounters and entanglements due to the mandates of instrumental reason and individual self-management. After all, the gallery spaces in which *The Visitors* have shown are neither battle fields nor stock markets, neither places of competitive upward mobility nor sites at which we as citizens of the Anthropocene are forced to come to terms with how humanity's efforts to control nature now produce man-made disasters we no longer can control. The spaces for *The Visitors* instead are marked as places of aesthetic experience, as locations in which audiences in the mode of play can probe sensations, affects, resonances, and possible actions at odds with the logic of the everyday. In such places, to attend to exploding stars without need or fear to jump to immediate action means no more and no less than to test—at a sensory level—what it takes, feels, and means to be hospitable amid deeply inhospitable times. It is to tap into contrafactual models to sense the seemingly impossible: to intend non-intentionality; to will the end of our strategic willfulness; to open doors to unburden us from the very dwellings we have built to safeguard our selves from the strange and other, including the strange and other in us.

Sociologist Andreas Reckwitz describes Western societies of the twenty-first century as societies of singularization. The expectations of singularity exceed neoliberal pressures of self-management and self-optimization. Paradoxical as it may seem, they turn desires for unique and extraordinary self-performances into dominant social norms. As a result, everything has to live up to the call for individuation: how you work, where and how you dwell, what destinations you choose for your travels, how you shape and dress your body, what circle of friends you assemble around yourself, etc. "In the mode of singularization, life is not simply lived, it is *curated*. The late-modern subject performs her or his (purposively) special self before others who become the audience."[25] In its very demand to produce the self as exceptional, singularization not only transforms others into spectators, but increasingly erodes the ground to welcome the other without expecting some kind of reciprocity: performers need spectators like commodities need buying costumers. As it approaches atomization as a spectacle and market place of first rank, Reckwitz's curated self thus increasingly depletes the necessary resources to offer and accept gestures of hospitality—to let come, give space, connect to, and resonate with others and the unknown in defiance of

[25] Andreas Reckwitz, *Die Gesellschaft der Singularitäten: Zum Strukturwandel der Moderne* (Frankfurt am Main: Suhrkamp, 2017), 9.

transactional expectations. The more singular we become, the less we are able to recognize and attune ourselves to the incommensurability of others.

Kjartansson's *The Visitors* turns the logic of singularization against itself. It climbs on its ladders to move beyond it. The video ends with all members of the band first meeting in one room and then being joined by the rest of the crew for a rowdy procession down the lawn of Rokeby Farm—so rowdy that the camera fails to frame the assemblage of bodies in cinematographically sound ways (Figure 2.5). The singing continues, though the repetitive hums of "Once again I fall into my feminine ways" will become increasingly faint. Because eventually all this is shown on only one of the nine screens, the installation's visitors will find themselves drawn to one common area in the gallery space and thus asked to assemble as one loose body of onlookers and listeners, as much attending to what is presented to them on screen as to the strangers with whom they have previously shared the gallery. We are wrong, however, to think of this ending as a ritualistic holy communion, fusing the dispersed, fragmented, and singularized into a spurious fantasy—a deceptive simulation—of successful togetherness. Or as a march beating shape and form into the amorphous and chaotic. What is at stake in this final assembly instead is both to display the

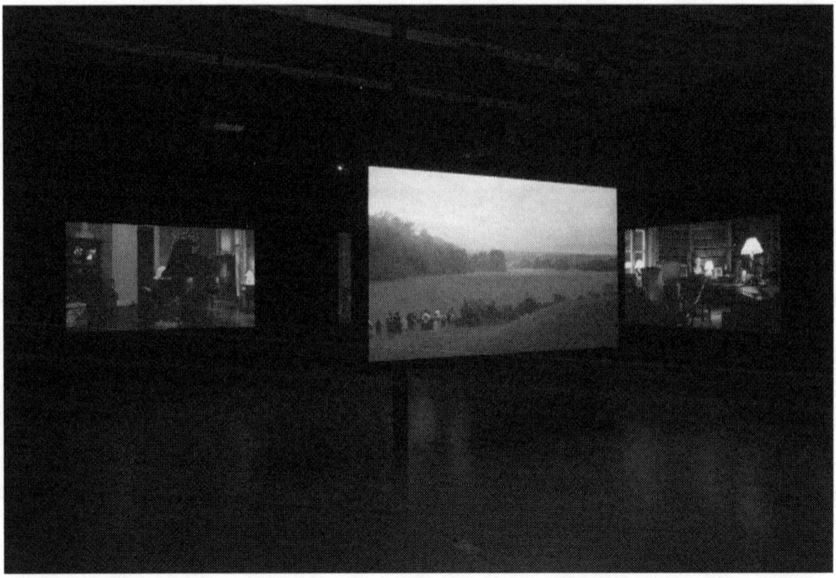

Figure 2.5 Ragnar Kjartansson: *The Visitors*, 2012. Installation view Luhring Augustine, New York, 2013.

dynamics of singularity as a social figure itself and to demonstrate its limits, the fact that singularity today is neither inevitable nor avoidable. Resonant attentiveness and collaboration here serve as a corrective to singularization's relentless logic of self-curation and transactional self-enclosure. It rubs against singularized society's categorical division between performers and audiences. It reminds us, in sensory terms, of what may reside beyond the industrious stage we have come to call the contemporary. In the (remaining) heart of our singularized existence *The Visitors*' final tableaux invites us to find what it might take to encounter companies of strangers, not as a mere audiences or competitors, but as vibrant confederations to build a future of alterity grounded in care and compassion.

"As a figure that cuts in, the stranger is a vibrancy that may disturb, but in doing so may also unwittingly fulfill our desires for cultural diversity—for what we may secretly desire: the wish to be disturbed, that is, to love and be loved: a love at first hearing. *This other that pierces.*"[26] Hospitality toward strangers neither requires friendship nor love. In essence, it simply—as Immanuel Kant argued in his *Perpetual Peace* of 1795—acknowledges a stranger's (cosmopolitan) right not to be treated as enemy when arriving in the land of another. It does not grant a right to stay forever, but it wants to allow unknown strangers to approach and enter one's land or house without hostility or fear, without transactional expectations and adversarial postures of exclusion or enclosure. If hospitality required friendship or love, it would fall short of its own universalist ambitions, its categorical insistence to welcome the foreign and other for the sake of their very foreignness and otherness.

Whether you are a Kantian or not, however, no formal definition of hospitality can preclude the possibility that friendship and love may yield from acts and experiences of hospitality after all. Nor does it mandate that moral commitments to being hospitable to strangers may not secretly express or encode our desire to love and be loved, to resonate with and be attached to the lives of others, to be decentered and disturbed by vibrancies that exceed our conceptual maps and understanding. Unlike Kant, who thrives at considering hospitable acts as cold moral obligations, Kjartansson makes us think of hospitality as a pink rose in the glittery frost. *The Visitors* tears into its listeners' lives (and so often produces tears in the eyes of its viewers) because it abrogates the deals of our

[26] Brandon LaBelle, *Sonic Agency: Sound and Emergent Forms of Resistance* (London: Goldsmiths Press, 2018), 72.

deeply inhospitable times, our era of curated singularity. In the very gesture of this negation, *The Visitors* reclaims hospitality's often unspoken missive—that only those can expect the bliss of friendship and love who know how to *fall into* possible entanglements and, with requisite patience and slowness, afford themselves to resonate with what transcends expectation.

3

Time

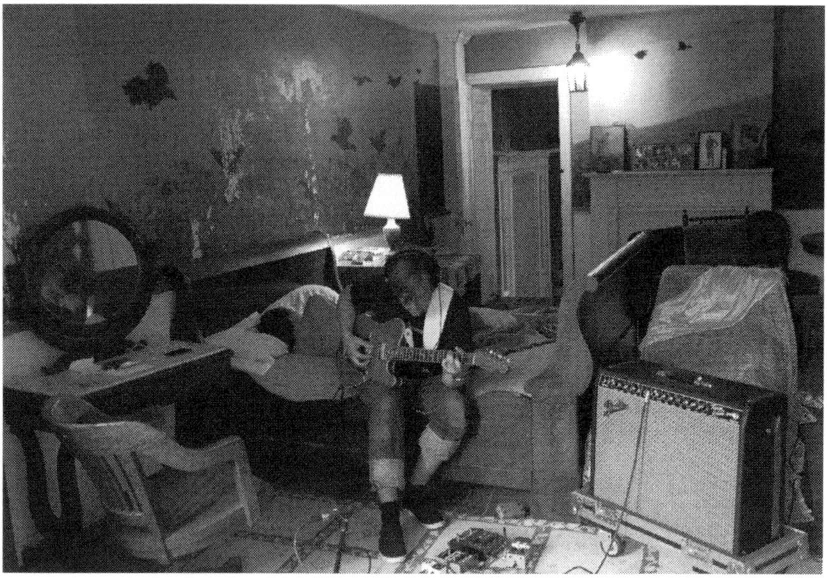

Figure 3.1 Ragnar Kjartansson: *The Visitors*, 2012 (Ólafur Jónsson). See also Plate 6.

On Slowness

The field of screen-based installation, performance, and sound art has witnessed remarkable forays into the slow and durational over the last decades. In *The Artist Is Present* (2010), Marina Abramović—endurance artist par excellence—sat for two and a half months in utter silence at a wooden table in the MoMA in New York, awaiting more than 1,500 unknown museum visitors who took turns to face her unrelenting gaze. To stretch performance art beyond expectations, to take time and make time in the absence of both speech and electronic distractions, according to Abramović's ambition, was to open portals to resonate with the lives and emotions of complete strangers. Christian Marclay's *The Clock* (2010) clearly set out to push the durational commitments of spectators as much as those of museum curators. Time in *The Clock* flies fast, never ceases to march forward, is always literally on time and in synchrony. And yet, the work's 24-hour format, its tour de force through international cinema's cutaway shots of timepieces, and its intricate efforts using sound to smoothly align around 12,000 different film clips into one continuous timepiece, hoped to rework the fickle attention spans of the present. Max Richter's eight-hour *Sleep*, first released in 2015 and then performed to audiences during a night session in London that invited the listener to do what the work's title proposed, did everything in contemporary music's disposal to deflate time and emancipate the listener from the digital age's haste and agitation. Blending 1960s minimalism, ambient music, and new age idioms into one stream of barely moving sounds, *Sleep*'s slowness aspired no less than to lull resonant listeners into what may exceed our time's restless monetization of consciousness.[1] In their efforts to stretch time, however, Richter's eight hours, Marclay's twenty-four-hour format, and Abramović's two and half months all pale in the face of the legacy of John Cage's desire to emancipate time from the footprints of the human. Installed in 2001 in St. Burchardi church in Halberstadt, Germany, his 1987 musical piece *ASLSP (As Slow As Possible)* is designed for a duration of 639 years, its chords shifting at intervals somewhere between a few months and multiple years, its final vibrations to be played in and for a future that may no longer know of human civilization.

[1] In the words of one of its enthusiastic commentators, the performance served as an elaborate sleeping pill: "It's a place in which you can settle for a while, with or without a pillow, and emerge only when you are ready to rejoin the restive world." Grayson Haver Currin, "Max Richter: Sleep," *Pitchfork*, September 21, 2015, https://pitchfork.com/reviews/albums/20987-sleep/.

Art's ventures into the slow are often couched in the language of mindfulness, authenticity, and resistance: as a means to contain, dispute, or displace contemporary regimes of 24/7 in whose context, according to Jonathan Crary's scathing account, the "idea of long blocks of time spent exclusively as a spectator is outmoded."[2] Aesthetic slowness, it is hoped, reconnects us with something digital economies of attention threaten with extinction: contemplative attachments, empathy and emotional resonance, deep listening and fastidious seeing, patient respect for the formal integrity of aesthetic objects, and appreciation of the human and its limits rather than of machines and their built-in temporal protocols. However, while much of this is timely and well-meaning, it often fails to answer a whole series of basic questions. How shall we identify the threshold of slowness in the first place? Does the slow and durational need contemporary speed and acceleration to define themselves as such, and if so don't they thus inadvertently reinforce the very conditions they defy? Is your sense of slowness really the same as mine and, similarly, is one culture's or social group's ideas of both speed and long blocks of time really identical with another's? And if not, doesn't the relativity of slowness resist any effort to couple aesthetic experiments with extended temporalities to any kind of politics, any kind of transformative, pedagogical, or therapeutic agenda, any kind of common platform at all?

Ragnar Kjartansson is no stranger in the world of the slow and the durational, of temporal experiments and repetitive delays that dare the temporal commitments of performers, curators, and audiences. In his 2007 single-channel video *God* (Figure 3.2), Kjartansson intones the phrase "Sorrow conquers happiness" over and over again for thirty minutes, his body occupying the space in front of an orchestra of eleven musicians and in a room draped with pink curtains, his hips swaying like those of a mid-twentieth-century entertainer. Ever-so slight changes in tone, volume, and intonation challenge the audience either to sink into the performance's iterative flow of sounds or simply leave the gallery asap. Kjartansson's contribution to the 2009 Venice Biennale had the artist inhabit a studio—Iceland's official pavilion—for the entire duration of the show, drink beer and smoke cigarettes, and paint 144 paintings (one for each day) of fellow artist Páll Haukur Björnsson while the recorded sounds and sights of a country music performance called *The End* could be continually witnessed in another room of the palazzo. The single-channel video *Song* (2011) featured three female singers singing lines of a 1954 poem by Alan Ginsberg—"The weight of the

[2] Jonathan Crary, *24/7: Late Capitalism and the Ends of Sleep* (London: Verso, 2013), 53.

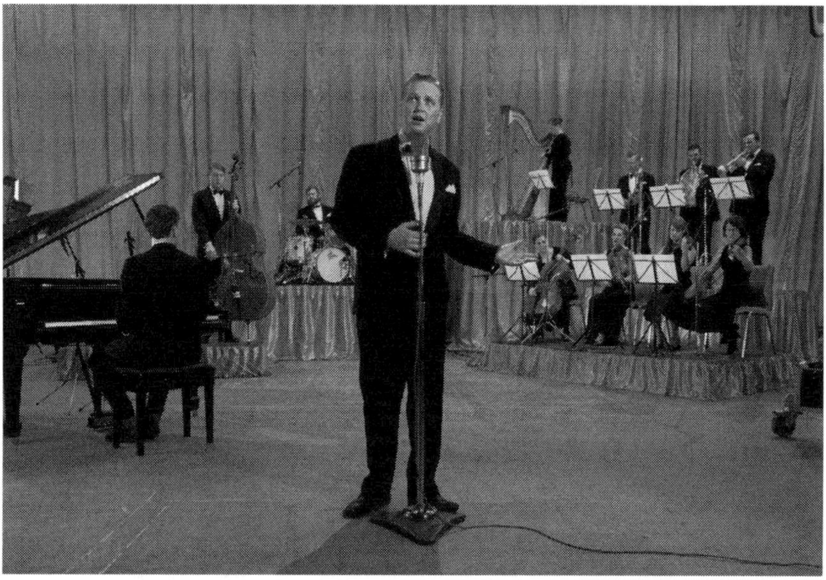

Figure 3.2 Ragnar Kjartansson: *God*, 2007.

world / is love . . . / No rest / without love, / no sleep / without dreams / of love"—for six hours amid Carnegie Museum of Art's impressive Hall of Mirrors. *A Lot of Sorrow* (2013) captured a concert of Kjartansson with the band The National in which one of the band's songs, "Sorrow," was stretched without interruption over six hours. And *Woman in E* (Figure 3.3), among others, performed at Kjartansson's mid-career retrospective at the Hirshhorn Museum in Washington, DC in 2016, asked a cast of fourteen female performers in gold gowns to individually step onto a rotating pedestal in sequence and strum a single E-minor chord on an electric guitar every few minutes for the entirety of the exhibitions' twelve-week run.

Though the length of his performances and recorded work can pose considerable challenges for the artist, his collaborators, and his audiences, Kjartansson does not think of himself explicitly as an endurance artist—a practitioner pushing the limits of what bodies and minds can take and in so doing pushing the limits of art itself.[3] Neither Abramović's biopolitical masochism,[4] nor Richter's invitations to rest soundly, describe Kjartansson's signature. What may

[3] "Sorrow on Repeat: Ragnar Kjartansson on Endurance Art," CBC Radio, January 20, 2015.
[4] Jaime Brunton, "Biopolitical Masochism in Marina Abramović's *The Artist Is Present*," *Camera Obscura*, 32.1 (2017): 63–91.

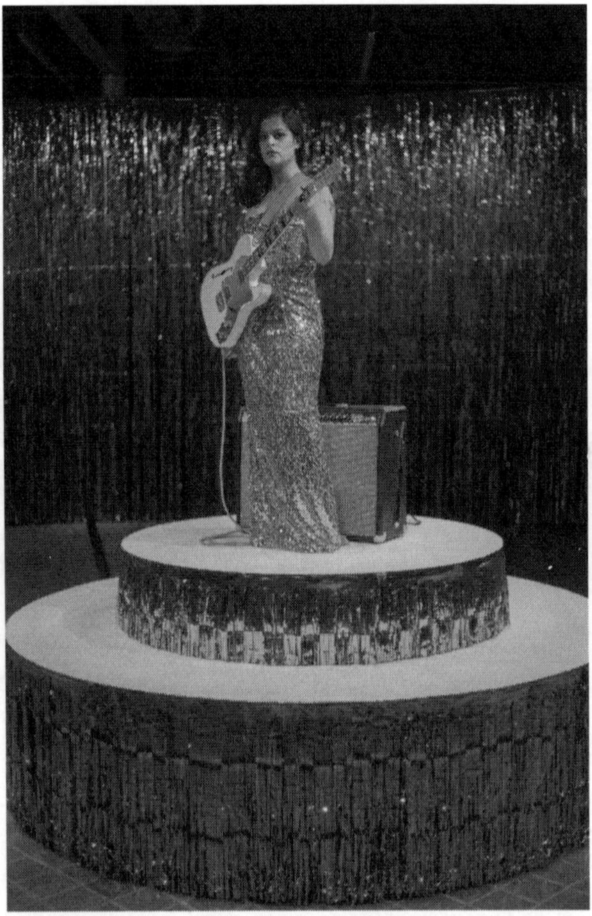

Figure 3.3 Ragnar Kjartansson: *Woman in E*, 2016.

define the provocative slowness of his work, instead, is the figure of iteration, Kjartansson's consistent aesthetic of repetition. Amid much of the rest of his work, the 64-minute loop of *The Visitors* may almost appear short and hasty. Yet the work and its formal structure ask tough questions about the role of distended temporalities and decelerated times in contemporary culture, as much as his play with repetitive elements—his refusal of linear progress—urges us to rethink what we expect of and how we should conceptualize slowness. Many may leave the installation after a few minutes, but many may also forget about the passing of time altogether. Whatever counts as Kjartansson's slowness is neither about merely inverting the speed of our age (and thereby dialectically reaffirming it) nor about discovering the slow and sluggish as membranes to return us to the presumable bliss of the local, the non-mediated, the authentic and unstaged. Rather, in its very play with repetition and difference, non-linear time and the durational, *The Visitors* offers a paradigmatic example of slowness as a mode of experiencing time in all its heterogeneity and multiplicity, and as a critical medium to establish, test, and recalibrate the wonders of resonant interactions between different agents in the world. True hospitality reveals itself, not in the short instant of the stranger's arrival, but several days into their stay: how long are we willing to bear with them? As showcased in *The Visitors*, resonance likewise needs and takes time to do its work.

Chanting for Enchantment

Refrains stand like lighthouses at the shore. Songs, singers, and listeners look out for them amid their journeys on the sea. They offer points of orientation and direction, their unflinching stability allowing that which moves to recognize, map, and indulge in its own movements. Like a lighthouse's flashing beam, refrains showcase repetition as a mechanism to keep the boats of individual stanzas or verses in motion. Though the words' etymologies are different, refrains refrain from breaking with circular time; they abstain, withhold, or desist from not doing the always-same so as to enable musicians and audiences alike to make the best out of their passage through time and space. Refrains hold back time in order to enable it; they curb temporal openness to facilitate its fabrication. Refrains, then, ground music's vibrations to allow a song's rhythms to venture out into and safely return from the pleasures of possible chaos, the storms of affect, the flows, waves, turmoil, and adventures of rolling seas. Or in Gilles Deleuze and Felix Guattari's somewhat cryptic formulation:

> The role of the refrain has often been emphasized: it is territorial, a territorial assemblage. Bird songs: the bird sings to mark its territory. The Greek modes and Hindu rhythms are themselves territorial, provincial, regional. The refrain may assume other functions, amorous, professional or social, liturgical or cosmic: it always carries earth with it; it has a land (sometimes a spiritual land) as its concomitant; it has an essential relation to a Natal, a Native.[5]

[5] Gilles Deleuze and Felix Guattari, *A Thousand Plateus: Capitalism and Schizophrenia,* trans. Brian Massumi (Minneapolis: University of Minnesota Press, 1987), 312.

Though Kjartansson is a master of deflecting the drive of teleological time, his refrains are of a different order, including and perhaps most prominently *The Visitors*' "Once again I fall into / My feminine ways." His refrains are neither lighthouses nor native land, neither rhythmic bird songs marking space nor flashing beacons ordering time. They neither orient the listener nor do they serve as promissory notes for safe returns. They repeatedly flash up amid the rolling of the musical sea, their location ever shifting, their rhythms taking on ever-different patterns. No sailor should or could ever trust them. Perhaps we in fact do well not to think of them as refrains at all. For if time repeats over and over again, even with small differences, how can we in any meaningful ways speak of the presence of refrains in the first place? If refrains are everywhere and deliberately forfeit their stability and stillness, they in some sense are nowhere at all.

In pushing refrains beyond their bounds and territories, copious repetition in Kjartansson's work aspires to no less than the logic of the chant. Sung or shouted, chants typically repeat rhythmical phrases in the hopes of making multiple voices resonate in unison or of endowing an individual's voice with the capacity to transcend itself, to break the presumed confines of subjectivity, and to thereby connect to something higher, larger, noumenal. St. Augustine chanted his prayers over and over again in order to direct his perception away from earthly matter and entirely focus his thoughts onto the spiritual presence of God. To chant was to subtract himself from material distractions and sing his faith into being, his attachment to the divine. Soccer fans chant during a game to mark territories and fire up the action as if there was no difference between the 50,000 bodies in the stands and the players on the pitch. These chants want to bind minds, voices, and bodies into one dynamic; they reverberate through the stadium to carry the ball to its goal (or keep it from doing so). "You'll never walk alone."

Chants are techniques of enchantment, or of re-enchantment as some might want to argue at the end of an age critical theorists named the modern one of total disenchantment. Though chants come in many shapes and sizes, what they all share is to offer a method of sharpening some form of attentiveness, of focus, whether such sharpening is to lead the chanter away from things sensuous, material, and aesthetic, or whether it is to entangle the chanter into the sensible world in joyful excess of ordinary understandings of cause and effect. Chants claim that attention not only matters, but that it needs to be trained and cultivated in order to recuperate the experience of awe and wonder. What, then, is the nature, the drive, the aspiration, the focus of Kjartansson's chants, of him declaring over and over again that it is time for sorrow to conquer happiness and for us to fall into our feminine ways? What kind of enchantment does the enduring power of his chants seek to generate?

To answer these questions, we need to dig a bit deeper. Chant, writes Thomas Moore, "is modal music, which means that it doesn't have the powerful drive that much of modern music has to arrive at a final harmonious destination."[6] To which Jane Bennett adds: chants

> provide sensory access into the cosmological (i.e., energetic and rumbling) dimension of things. To sing or hum or otherwise enter into a musical refrain is to get into a groove—your torso sways side to side, your chin nods north and south, your toes tap up and own, your chest dips down and out. But what is distinctive about this groove is that it is not wholly smooth; each of its repetitions never exactly repeats its predecessors. One's postural (spiral) repetition mimic the audible repetition of the refrain, which itself reenacts the (spiral) motility of cosmos.[7]

As a structure of attention that requires technique and cultivation, the chant's groove—according to Bennett—animates and entertains feelings of enchantment often presumed to be lost in the wake of Enlightenment reason and industrial rationality. Chant today has the power to reveal that enchantment is not entirely gone, though most of what may count as chant in the modern world is different from St. Augustine's fifth-century efforts to sing himself beyond sensuous attachments. For Bennett, chant's most important modern focus is on vibratory entanglements that cut across presumed divisions of mind and matter, subject and object, the human and the nonhuman. In their very repetitive swirling and swaying, chants can foster feelings of affirmative connectedness—of resonance—that approach the sensible as a foothold of possible futures that undo past histories of repression and oppression. Chants train sensibilities that prepare our senses for experiences of surprise and wonder, that is, for experiences that exceed anticipation and preexisting concepts, narratives, and meanings, that rupture the fabrics of time without—like trauma and shock—overwhelming or petrifying the senses.[8] The chant's enchantment enables momentary impressions

> that the natural and cultural worlds offer gifts and, in so doing, remind us that it is good to be alive. This sense of fullness—what the Epicureans talked about

6 Thomas Moore, *The Re-Enchantment of Everyday Life* (New York: HarperCollins, 1996), 108.

7 Jane Bennett, *The Enchantment of Modern Life: Attachments, Crossings, and Ethics* (Princeton: Princeton University Press, 2001), 166.

8 For more on the aesthetics of wonder, see Philip Fisher, *Wonder, the Rainbow, and the Aesthetics of Rare Experiences* (Cambridge, MA: Harvard University Press, 1998); and Lutz Koepnick, *The Long Take: Art Cinema and the Wondrous* (Minneapolis: University of Minnesota Press, 2017).

> in terms of ataraxy (contentment with existence)—encourages the finite human animal, in turn, to give away some of its own time and effort on behalf of other creatures. A sensibility attuned to moments of enchantment is no guarantee that this will happen, but it does make it more possible. Any sensibility is an orchestrated arrangement of affections, but affective energies are unruly and protean forces and tend to wander from the musical score. Thus, the link between them and an ethical sensibility is tenuous and unstable and requires repeated acts of discipline and retuning.[9]

Chants embed us in iterative time in order to open secret doors for futures different from calamitous pasts. Their modal form defies final destinations so as to make us hold on to the promise of a future in which fear may no longer need to drive our relations to the matter of the world, including the world and other inside ourselves.

At first impression, the performers of *The Visitors* appear everything but eager to engage in chant's slow orchestration of affect, let alone ready for repeated acts of (self-)discipline. What we initially seem to witness is a bunch of hipsters and hippies, systematically flouting a chanter's rigor and self-restrain, his or her desire to delay desire and gratification. Kjartansson lounges naked in a bathtub; Davíð Þór Jónsson consumes big cigars and various drinks while playing the piano; the flowing attire of both Valtýsdóttir twins communicates utter repose rather than a meticulous work ethic; Ólafur Jónsson performs in front of a bed featuring a sleeping female body. Ataraxy, the images of the installation tell us, is on the performers' side already; Epicurean fullness has arrived long before the first sounds vibrate across the old mansion and we as visitors of the installation focus our eyes and ears on the performance.

And yet, what will develop over the next sixty-four minutes features nothing other than the work, the training and cultivation, it takes to sing and play music in unison, to bridge physical separation without collapsing spatial distance, to put to work the channels of acoustic media so as to achieve what Bennett calls enchantment and this book understands as a wondrous instance of resonance. To repeat various sections of *Feminine Ways* over and over again serves the band's performers as a technique to train their attention onto each other: to search for, probe, secure, and lose again possible channels of communication and togetherness. Repetition in *The Visitors* here offers a medium of inquiry and receptivity, of pushing against the confines of singularity. Like the oscillations

[9] Bennett, *The Enchantment of Modern Life*, 156–7.

of a chant, the song's modal stretching of time, its denial to tell a forward story and arc toward a final destination, affords each of the musicians with necessary resources to listen for and lean into each other—to fall into each other's groove and discover moments of synchronicity and coexistence, however ephemeral they may be. Acts of repetition here, in other words, unlock a space to experiment with different approaches toward a joyful commons whose resounding ataraxy is doomed to vanish in no time again. Though the song—like a chant—will not really develop its musical material in ways satisfying musical experts, the performance as a whole—like a chant—aspires to a retuning of social sensibilities and the networks of community. Contrary to our first visual impression, the performance distends time, not to challenge our patience or provoke our endurance, but to remind us of the treat of being alive. It rigorously explores the slowness of iterative sound as a medium to unwind the rigor necessary to navigate a present stacked with relentless dynamics of power and distraction, self-management and coercion.

As both a process and a product of cultivating attention, chants are often associated with meditative restraint and subdued forms of conduct. Soccer fans, eager to carry the ball like modern magicians with their voices, will tell you a different story. And so do the members of the cast of *The Visitors* when they enter the second loop of singing "There are stars exploding / And there is nothing you can do" more than thirty minutes into the performance. The deliberate probing of coexistence here turns into outright shouting, a rowdy crescendo of voices whose unity resides in defying harmony and synchronicity. The graphic score identifies the section in question as "gospel." It comes as close to a moment of climax and contraction modular music might ever get. Kjartan Sveinsson stands next to Jónsson's piano, their physical proximity helping both to amplify their expressive energy. Kristin Anna Valtýsdóttir cups her hands over her earphones and powers up her voice by shaking her legs passionately. Shahzad Ismaily, whose preferred position throughout the rest of the performance is a leather armchair, now stands up and swings his entire body while playing his bass guitar. There is not much you can do when facing celestial catastrophes that have happened many light years ago. You can go inward and mourn. Or you can shout. Shout with others. Repeat your shouting over and over again. Seek to resonate with and echo the shouting of others. Scream together and in screaming together share time with others, well aware that such affective energies cannot but be unruly and make you wander away from traditional ideals of individual sovereignty. Scream at the top of your lungs to abandon control over your voice and precisely thus remind yourself of how good it is to be alive. How good it is to connect to and be touched by what is not you.

Things calm down again after this crescendo. The song reboots and navigates calmer waters. The singers are far from done, though. The probing search for ataraxy, for the joy of touching upon and being touched by the vibrations of others, continues. The chant carries on its paradoxical quest for the presumed effortlessness of resonance. Nineteenth-century French writer Marie-Henri Beyle, better known by his pen name Stendhal, once famously considered art and beauty a *promesse du bonheur*.[10] Whether the band members sing or shout, whether critics think of Kjartansson's installation as art or not, whether audiences swing along the band's at once rigorous and unruly chanting or not: what animates the logic—the ethics—of repetition in *The Visitors* is nothing less than the effort to embrace resonant matter as a promise of happiness, of enchantment. *The Visitors*' slowing down of time is not about enduring the present. It is about hope, however melancholic and faint it may come our way. It is about holding on to the promise that human animals may experience some sense of fullness once they learn how to abandon their desire to own time and are ready to let in and attach to what exceeds existing concepts, expectations, narratives, and systems of order.

Against Conducting

Futurist speed aficionados of the early twentieth century had good reasons to love cars and airplanes. These vehicles not only promised unprecedent velocities but also promoted drivers or pilots to intoxicating positions of control and mastery. Speed—the act of bolting forward along a linear axis—served as a medium of power and sovereignty, including the power to put one's own sovereignty on the line.[11] In this, modernist advocates of speed often modeled their posture on the nineteenth-century figure of the virtuoso performer: violinists or pianists such as Nicolò Paganini or Franz Liszt whose fame rested as much on their expressive gestures as it did on their hallucinating coupling of dexterity and speed, on their

[10] Stendhal, *De L'Amour* (Paris: Gallimard, 1980), 58–9.

[11] Enda Duffy, *The Speed Handbook: Velocity, Pleasure, Modernism* (Durham: Duke University Press, 2009); Lutz Koepnick, *On Slowness: Toward an Aesthetic of the Contemporary* (New York: Columbia University Press, 2014), 15–52.

skill to play instruments at humanly unthinkable levels and drive sound forward as if their instruments played themselves.[12]

The invocation of slowness in contemporary art often presents itself as a medium to renounce such couplings and crossings. A recurrent theme of what counts itself as slow art, as durational and endurance practice, seeks to eliminate the role of author, composer, maker, or performer from the picture altogether. In doing so, it hopes to undercut speed's presumed leanings toward forms of subjectivity that either undergird muscular gestures of control or result in a defensive armoring of the body against accelerating rates of stimuli. Consider Brian Eno's *77 Million Paintings* (2006) (Figure 3.4), a digital art work that has been shown as a single-channel or multichannel installation in galleries and museums across the global ever since it was projected onto the sails of the Sydney Opera House in 2009. *77 Million Paintings* draws on 296 original works that during any screening are combined into constellations of four, slowly and asynchronously dissolving into ever-new combinations over extended periods

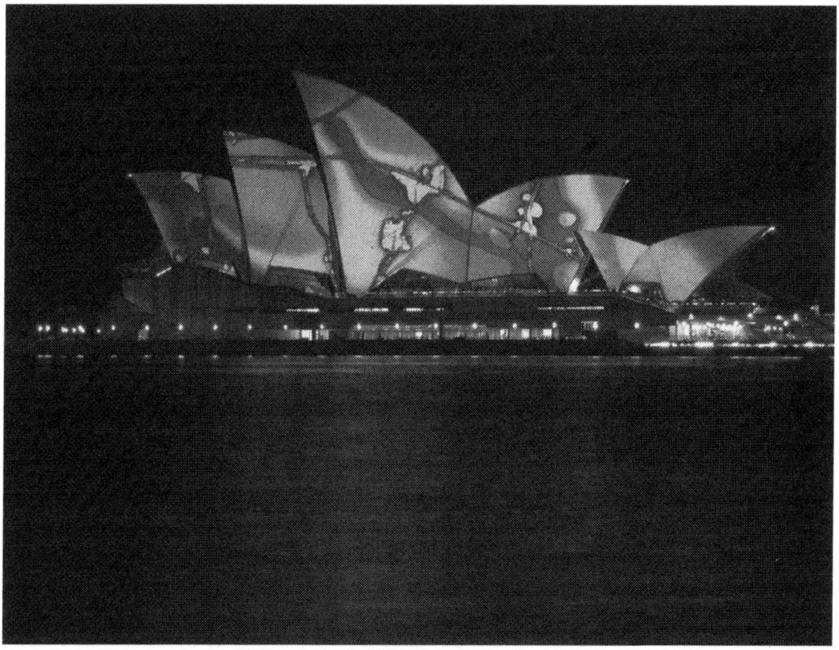

Figure 3.4 Brian Eno: *77 Million Paintings*, 2006.

[12] Paul Metzner, *Crescendo of the Virtuoso: Spectacle, Skill, and Self-Promotion in Paris during the Age of Revolution* (Berkeley: University of California Press, 1998), 113–59.

of time. Computer software randomly selects individual elements, determines the duration of their display, and thus enables what is, from the perspective of human perception, a quasi-infinite number of possible variations. Computer algorithms also generate randomized musical sounds meant not simply to decorate the stream of images, but to offer ambient soundscapes that envelope viewers and invites them to relax into the absence of everyday strains of time. *77 Million Paintings* wants both very little and everything from its audience. The processing speed of computer chips here permits temporal experiences in whose context nothing ever seems to repeat itself, yet whatever is new and different arrives so smoothly that viewers will feel comfortable to put perceptual shields to rest and simply become absorbed into what transcends human authorship and intention. Twenty-first century drivers and pilots will no doubt consider the speed of Marinetti's cars and airplanes as sluggish and glacial—as something no longer able to fascinate anymore. In emancipating time from the nervous tensions of the human, however, *77 Million Paintings* is designed to lull even the most nervous Italian futurist into comforting contemplativeness, perhaps even sound sleep.

Durational works such as Cage's *ASLAP* and Richter's *Sleep* certainly anticipate and echo the principal proposition of *77 Million Paintings*: slow time does not only soften the edginess of contemporary speed addicts; it happens when you pass on the work of producing artistic works to algorithms, machines, randomizers, and computational processes. In Cage, Eno, Richter and others, sound's ability to surround and embed the listening subject, its defiance of vision's stress on frontality and directionality, and its invisibility and presumed privileging of affect and entanglement over intellect and distance situate the audible as a perfect realm of slowness, a medium to deflate the fantasies of combative power associated with modern speed. Sound here is slowness's natural-born ally. It figures as a perfect training ground to reeducate the entire human sensorium.

Kjartansson's politics of time complicates these aspirations. For the modular character of *The Visitors*—the fact that neither the installation's sounds nor its images really want to go anywhere—plays a crucial role, not to evacuate human subjectivity, but on the contrary to secure a meeting ground for diverse subjectivities to attach forces without submitting their autonomy to one command, plan, or effort. Nothing in *The Visitors* unfolds automatic, nothing aims for the comfort and consolation that algorithms might offer to restless human souls. The full import of its conception of slowness instead is best understood when mapped against the nineteenth-century idea and ideology of symphonic music, as embodied in the charismatic figure of the conductor. As Christopher Small recalls:

> For the modern conception of conducting to take hold—the conductor in charge of every detail of the performance and the sounds blended into a unified texture that is directed toward outside listeners—it was necessary that the players' musical autonomy and power of independent action be abolished. A musician in his role as conductor cannot exert that kind of control over an orchestra of improvising musicians, nor can he in his role as composer be sure that the sounds that are made by the orchestra are the ones he imagined, as is required by works of music that are intended for listening rather than participating.[13]

Enthusiasts of Western art music are often partial to argue that a conductor's authority is key to ensure that a musical performance does not become stale and lose its edge. It requires scrupulous conductors, they insist, to maintain focus and ensure that individual instruments do not slip out of sync. Seated in a bathtub, Kjartansson wants to liquify upfront any expectations audiences might have about the role of the conductor and how composers might use that role to lead and beat performing musicians into submission. Process is *The Visitors*' product. Slowness here provides a laboratory space to negotiate the autonomy of musical voices and take different approaches to blending sounds into some unified texture. The song's repetitions offer a provisional platform on which each musician—with no conductor in sight and no composer or virtuoso taking command—engages in an ongoing give and take, leading and being led by one another to a few wondrous moments of unison.

In a recent essay, Jan Slaby has proposed the concept of relational affect in order to study how affect structures interpersonal and transpersonal relationships and traverses between bodies and subjects whose borders are fluid. Relational affectivity describes dynamic forms of rhythmic coordination, bodily entrainment, and emotional resonance that not only render selves, habits, and orientations porous, but indicates the extent to which the social is deeply inscribed in what we typically take for granted as the individual or personal dimension of different actions and emotional states: "The intuition behind this is that even in the domain of feeling—long thought to be a sphere of paramount individuality—the social is making its presence felt from the start. In countless

[13] Small, *Musicking*, 83.

situations of our being affected, what we feel is not fully 'our own' but from the outset part of a relational tangle that exceeds our individual reach."[14] Relational affectivity allows us to experience different states of absorption, immersion, and rapture as important segments of our affective lives. As importantly, however, it reminds us of the extent to which different modes of entanglement precede what we consider subjectivity and individuation in the first place—the extent to which affect and emotion are social in the first place.

To take time and test its flows through improvisational gestures, in *The Visitors*, means to produce, feature, and host time as a vibrant space of relational affectivity in Slaby's sense. Speed consumes and kills time contrary to its own promises, just as much as charismatic conducting synchronizes different temporalities, rhythms, and speeds at the cost of obscuring what brings them together and will drive them apart again. *The Visitors*' slowness aspires to much more than simply inverting the speed of the contemporary. In contrast to both the nineteenth-century ideology of the conductor and to populist visions of decisive political leadership, *The Visitors* embraces the slow and modular as a condition for the possibility of reciprocity and non-hierarchical interaction, of immersion and entanglement. Like democratic deliberation, music making here needs time and patience to balance the agencies of different subjects. Slowness is the name for what it takes to generate temporalities in whose context diverse agents can suspend the asymmetries of autocratic control, train their attention on each other, attune their bodily rhythms, and precisely thereby investigate the promise of different experiences of resonance and wonder, however temporary they may be.

"For the listener and the performer," French philosopher and musicologist Vladimir Jankélévitch once wrote about the role of repetition and relational affectivity in music,

> the reprise is no less a renewal: hearing again, playing again, become modes whereby to discover, interminably, new relationships of subtle correspondences, beauty kept secret or hidden intentions. The polyphonic superimpositions of several independent voices that are nonetheless arranged in accord with another, the multivocal ambiguity that results, the innuendos and allusions that

14 Jan Slaby, "Relational Affect: Perspectives from Philosophy and Cultural Studies," *How to Do Things with Affects: Affective Triggers in Aesthetic Forms and Cultural Practices*, eds. Ernst van Alphen and Tomáš Jirsa (Leiden: Brill Rodopi, 2019), 72.

> accumulate in these superimposed levels, the unsaid things that they conceal like hidden treasures—in these no doubt there is a source of inexhaustible pleasure.[15]

It is tempting to think of *The Visitors*, in Jankélévitch's terms, as a treasure hunt—with the exception that its probing of polyphony and multivocality across different spaces never assumes the hastened temporality of a hunting exercise. Each reprise of "Once again I fall into / My feminine ways" and of "There are stars exploding / And there is nothing you can do" discovers new aspects and secrets, not about the musical material itself, but about what it means to share the space of music and superimpose different vibrations, rhythms, memories, and anticipations onto each other. Each repetition approaches and puts a different spin on how different voices, sounds, bodies, and materialities can resonate with each other without losing their difference.

Speed aficionados of the early twentieth century often embraced velocity as a crucible to melt the messy pluralism of the modern into lasting impressions of unity. Twenty-first century advocates of slow art and film often understand the slow as a source of meditative stillness and withdrawal, meant to reconnect the subject to something more authentic and lasting, to a form of unified temporality insulated from the speed and interruptions of the contemporary. Kjartansson's politics of time begs to differ from both. In his work, to repeat and be slow is to tune into and retune a present understood as a crossroad of multiple and often conflicting rhythms, narratives, vibrations, directions, and temporalities. Kjartansson's slowness does not shut down contingency. It is not troubled by the messy, complex, manifold, and unpredictable. It instead wants to sharpen our attention for experiencing the changing landscapes of the present in all their temporal multiplicity, mediation, and virtual openness. In this, the slowness of *The Visitors*, in spite of all its romantic melancholia, reminds us of the fact that nothing is inevitable and that everything could be different from how it is and has been.

Praising Blandness

It would be foolish to claim that repetitive elements are unique to the music of Kjartansson, Eno, Richter, and others; they are of course at work in all forms of

[15] Vladimir Jankélévitch, *Music and the Ineffable*, trans. Carolyn Abbate (1961; Princeton: Princeton University Press, 2003), 24–5.

music as we know them, whether classical, romantic, pop, or punk. We would be equally mistaken to think that music could ever entirely break away from organizing musical time as a pattern of dramatic build-ups and climactic releases or as a dialectical development presupposing highly attentive acts of listening. The minimalism of the 1960s—the experiments of LaMonte Young, Terry Riley, Steve Reich, and Philip Glass—tried hard to deflate musical time through iterative structures. Minimalism may have erred about its own ambitions, but it still defines one among other origins for the twenty-first century's strike against vectoral movements, narrative arcs, and climactic drama. "In much Minimal music," as Edward Strickland has summarized, "overt and immediately audible repetition of simple, even simplistic, material is the predominant structural principle. . . . Minimalism exposes the components of its medium in skeletal form."[16]

It is worthwhile to explore the minimalist echoes in Kjartansson's aesthetic of repetition, chant, delay, and slowness in greater detail. Let me chose a somewhat other direction here, however, to conclude my thoughts on the relation of slowness and resonance in *The Visitors*, inspired by what both musical minimalism and Kjartansson's iterative impulse have often been accused of: to savor, almost pathologically, the dull and bland. This shift in perspective will allow us not only to complicate some of the one-dimensional aspects of recent writing on slowness but also draw attention to the visual dimension of *The Visitors* and how the installation's iterative sounds echo with its nine seemingly unmoving screen images.

The work of French philosopher and sinologist François Jullien is the go-to place to think about blandness, even if the principal aim of his seminal 1991 book *Eloge de la fadeur* is to identify a cornerstone of traditional Chinese aesthetic practices.[17] This is not the place to get involved in debates about the Orientalist underpinnings of Jullien's argument. What instead matters is to explore the concept's ability to illuminate what we see at work in Kjartansson's repetitions and bending of time. "First," Jullien begins his treatise, "one accepts the paradox: that to honor the bland—to value the flavorless rather than the flavorful—runs

[16] Edward Strickland, *Minimalism: Origins* (Bloomington: Indiana University Press, 2000), 13. See also, among many other, Robert Fink, *Repeating Ourselves: American Minimal Music as Cultural Practice* (Berkeley: University of California Press, 2005); and more recently, Kyle Chayka, *The Longing for Less: Living with Minimalism* (New York: Bloomsbury, 2020).

[17] François Jullien, *Éloge de la fadeur. À partir de la pensée et de l'esthétique de la Chine* (Paris: Philippe Picquier, 1991); *In Praise of Blandness: Proceeding from Chinese Thought and Aesthetics*, trans. Paula M. Varsano (New York: Zone Books, 2008).

counter to our most spontaneous judgment (and elicits a certain pleasure in this contradicting common sense). But in Chinese culture, the bland is recognized as a positive quality—in a class, in fact, with the 'Center' (*zhong*) and the 'Root' (*ben*)."[18] In Jullien's account, blandness—the absence of flavor; a lack of marked distinctions and hierarchies of taste; a state of undifferentiated continuity and contiguity—constitutes the foundations of all things, a sense of plenitude from which particular qualities, defining values, and clear distinctions may emerge in the first place. In the realm of painting, as most paradigmatically embodied in the work of landscape artist Ni Zan (1301–74) (Figure 3.5), the pursuit of blandness led to visual compositions drained of all opacity; a narrowing of color ranges; a certain flattening of the difference between foregrounds, middle-grounds, and backgrounds allowing the eye to travel unhurriedly and evenly across the scroll; and a use of brushstrokes, not to record traces and distinctions but to allow marks to become one with forms to be represented. Nothing in Ni Zan's images of blandness agitates the viewer, strives to compel, direct, arrest, seduce, or fix the viewer's gaze. Everything, instead, remains muted and quiet, allowing the viewer to remain at once detached, curious, and open to the unexpected.

Though dedicated to local detail and a clear precision of strokes, Ni Zan's aesthetic of blandness captured a presumed balance and harmony from which other images and landscapes could emerge in the first place. The bland, for him, described a state of pure potentiality, the world's well of ongoing becoming and transformation. In order to have and distinguish different tastes, to make and mark distinctions, to map, navigate, and involve oneself in the world, one first had to experience the simplicity of all things prior to all possible differentiation, a state of detachment produced, not by unperceptive boredom or something merely pleasing to the eye, but on the contrary by beholding the underlying familiarity of all things, including the continuity between viewer and viewed. Bland images generated a form of spectatorship able to freely wander across and into the image space; a mode of durational looking that recognized the reciprocal relationships between viewer and viewed and shunned the agitation of narrative emplotment and teleology.

Its extravagant gestures and bohemian appeals notwithstanding, *The Visitors* at heart offers a twenty-first century excursion into blandness. Its sounds do not really go anywhere, not to bore performers and listeners into dullness, but on the contrary to investigate the seemingly dull and boring as a ground of open-

[18] Jullien, *In Praise of Blandness*, 27.

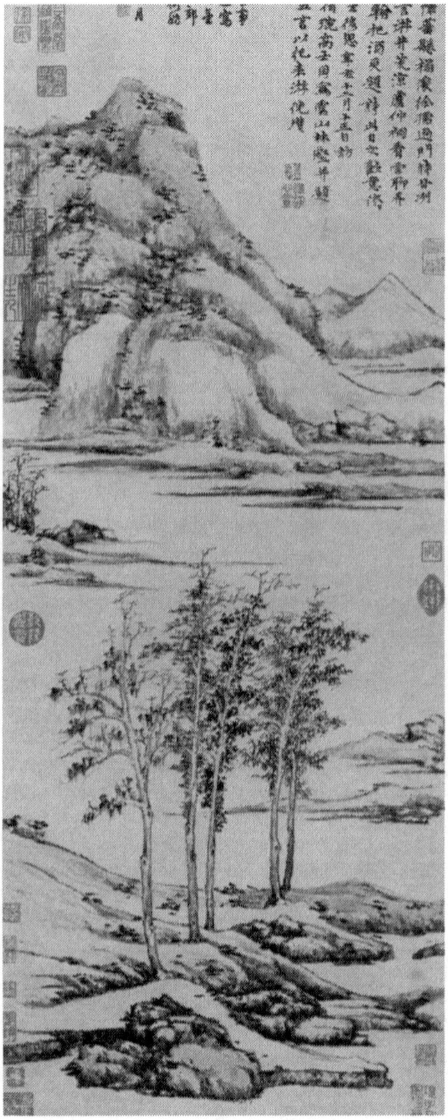

Figure 3.5 Ni Zan, *Trees in a River Valley in Yü shani*, 1371.

ended emergence and becoming, of unpredictable and ever-new instantiations of wonder and resonance, the wonder of resonance. Rather than bind the listener's attention through dramatic build-ups and releases, the performance of *Feminine Ways* opens a space in which listeners can investigate at a sensory level what it might mean and take to be attentive. As the music largely refrains from narrative entrainment and strong teleological arcs, it invites us to wander in and out of its audible texture so as to attend to minute differences among the same and similarities amid differences. To be sure, Ni Zan's metaphysics of primordial unity here no longer resides in the background. But blandness's aspiration to quell the agitation of the aspirational, to disentangle the perceiving subject from overstimulation so as to probe our state of entanglement, to formally disengage perception from temporal vectors in the hopes of allowing listeners to freely listen on their own temporal terms—all this describes the project of *The Visitors* as well and defines the music's undeniable simplicity as a source of unbound potentiality.

The blandness of *The Visitors* music is complemented by the installation's bland cinematography. The cameras capturing the footage for eight of the nine screens are static throughout the entire loop. We witness the process by which a crew member turns the cameras on their tripods on and off, one by one. But the images they produce are marked as utterly mechanical, as distant and disengaged, as purely observational, thanks to the fact that no human operator attends to their work during the recording itself. While our eight performers do what they do, the frames of their images take on the stoicism of surveillance devices. Nothing seems to touch the cameras' unmoving long takes. No cuts, no pans, no zooms indicate any response to the profilmic. No blinking of the observational eyes. These cameras record for the sake of recording. Like good hosts, they take in whatever comes their way. As they remain detached from whatever moves within the frame, they abstain from guiding the viewer's view and thus enable our gaze to navigate freely across their images or wander from one screen to another at our own pace.

No artist in their right mind, one may think, wants to count as an artist of blandness. After all, contemporary art requires the spice of disruptive gestures to stand out amid the pressures of contemporariness. And yet, it appears fitting to think of both the visual and the acoustical choreography of *The Visitors* as adding curious chapters to the history of aesthetic blandness. The piece's chants, its staging of iterative time, its poetics of delayed destinations and improvisational under-determination: all these different elements of *The Visitors*, like Ni Zan's landscapes of the fourteenth century, mark the space of the aesthetic as a site whose very blandness enables the possibility for strong affects and unexpected attachments; all these aspects, as they translate Ni Zan's flattening of representational planes into the languages of twenty-first century time-based art, brush against our times' agitated hunt for the next big thing and recalibrate art's time as a hospitable ecology to explore wondrous resonances between viewer and viewed, listener and listened.

Neither the slow nor the bland can expect to meet wide-ranging critical appraisal today. At best they are often seen as affirmative in nature, at worst they are simply shelved as boring. Both views crucially miss how Kjartansson's aesthetic of repetition taps into forgotten sources of enchantment by negating 24/7's ruthless logic of disruption, and how it approaches boredom, not as a burden, a curse, or a sign of exhaustion, but as a platform to await and lean toward the coming of a fundamentally unknown and unknowable future. In face of the accelerating time tables and ever-shrinking geographies of 1920s modernity, Siegfried Kracauer once praised boredom—the ability to linger and wait—as a unique medium to sharpen one's attention toward the present's possible futures. Waiting, he argued, "signifies an openness—one that naturally must not in any way be confused with a relaxation of the forces of the soul directed toward ultimate things; rather, quite the contrary, it consists of tense activity and engaged self-preparation."[19] Whatever is bland and slow about *The Visitors'* arrangement of sounds and sights welcomes restless audiences to reacquaint themselves with the art of waiting. It disavows the present's reigning economies of attention and invites us to discover boredom as one of the last remaining repositories of enchantment and the wondrous.

[19] Siegfried Kracauer, "Those Who Wait," *The Mass Ornament: Weimar Essays*, trans., ed. and with an intro. Thomas Y. Levin (Cambridge, MA: Harvard University Press, 1995), 139.

4

Water

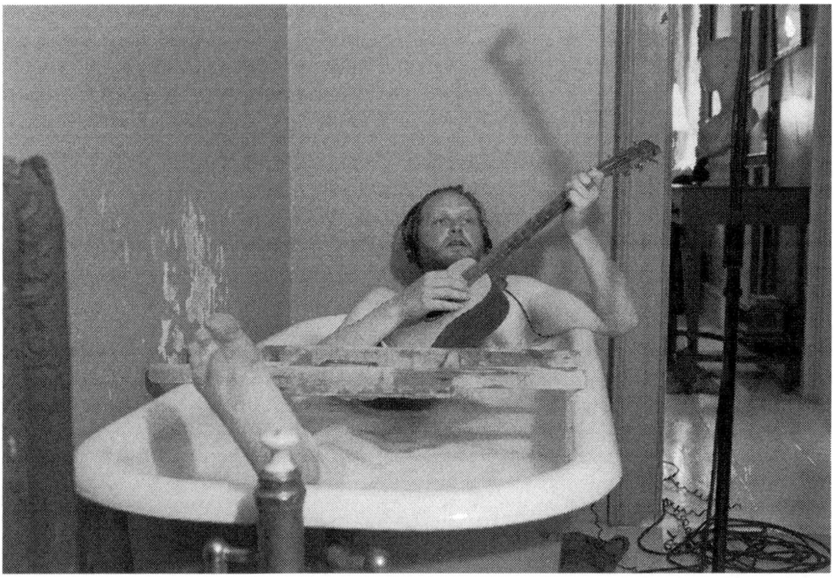

Figure 4.1 Ragnar Kjartansson: *The Visitors*, 2012 (Ragnar Kjartansson). See also Plate 3.

Ragnar's Bathtub I

Ragnar Kjartansson is certainly not the first artist transforming a bathtub into a site of aesthetic creativity and public visibility. A brief and admittedly patchy history of tubs in recent art, their iconicity and paradigmatic use, could read as given in the paragraphs that follow.

On April 30, 1945, the eve of the official announcement of Hitler's suicide, American photographer Lee Miller famously moved for three days into Hitler's Munich apartment and asked her colleague David E. Sherman to picture her in Hitler's bathtub (Figure 4.2). Each of these highly choreographed images shows Miller on the day she visited the concentration camp in Dachau, purposely wiping the camp's mud on the bathmat. They show her naked in what is a very

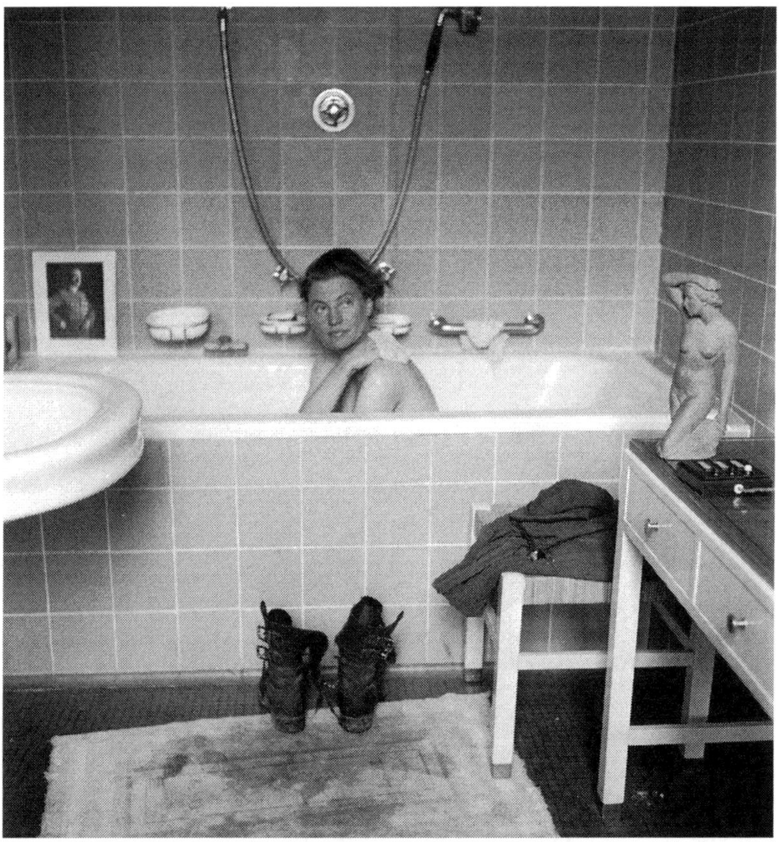

Figure 4.2 Lee Miller in Hitler's bathtub, by Lee Miller with David E. Scherman, 1945.

ordinarily looking tub, slightly posing, a small photographic portrait of Hitler in uniform sitting on the tub's rim on the left, a small sculpture of a nude female body on a table on the right. Miller's clothes are lying on an adjacent chair with her wrist watch placed on top of the pile, her military boots situated in front of the tub as if she had stepped right into the bath without allowing her bare feet touch the bathroom's floor at all. None of these pictures display Miller looking straight at and into the camera. Most of them have her gaze veer slightly to the left, in what could be read as introversion, coyness, or absentmindedness, and yet it is impossible not to see them as providing a clear symbol of Miller's anger and disturbance. All of them enact a curious scene of both physical and symbolic cleansing amid the very heart of fascism's darkness.

From the late 1978 to the early 1993, New York-based photographer Don Herron invited numerous art friends and celebrities such as Robert Mapplethorpe, Keith Haring, Peter Hujar, and Annie Sprinkle, along with Warhol Superstars like Holly Woodlawn and International Chrysis, to participate in his so-called Tub Series (Figure 4.3). His camera installed straight above the tub, Herron's images captured his subjects in at times highly stylized poses, at times no-nonsense, bare naked approaches. The intimacy of the tub here provided a site to perform and play with different configurations of identity, to appeal to and push against the New York's downtown cult of art stardom. In each shot, familiar faces could be seen in new and unknown ways. Each image offered glimpses of at once sensual and vulnerable bodies no matter the extent to which each of Herron's subjects situated themselves as objects to be recorded and seen by a larger public.

Referencing Jacques-Louis David's painting *Death of Marat* (1793), Thomas Demand's 1997 photograph *Badezimmer* (Bathroom) (Figure 4.4) shows the edge of a bathtub in a Geneva hotel room in which German politician Uwe Barschel—after a prolonged scandal over Watergate-like cross-party spying—was found dead by two reporters in 1987. Demand reconstructed the deed's scene with the help of his signature material and technique: paper and cardboard. Nothing in the physical model or the image indicates any trace of what may have happened in Geneva in 1987. The space is shown with dispassionate distance, its surfaces appear clinical and untouched. The bathtub reveals neither the presence of water nor of a dead body. In so doing, Demand not only reworked press imagery that circulated after Barschel's mysterious death, but employed the oblique sight of a bathtub as a medium to interrogate how we tell stories today in the realms of both the private and public in the first place.

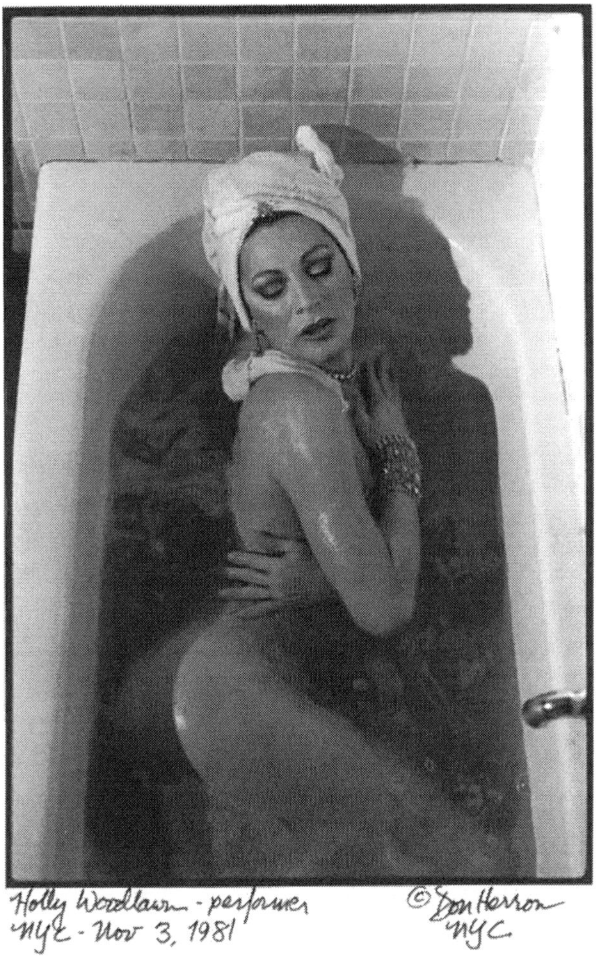

Figure 4.3 Don Herron: *Tub Shots—Holly Woodlawn,* 1981.

And then, there is of course Joseph Beuys's infamous *Badewanne* (1960) (Figure 4.5), scrubbed clean of its fat, cooper wire, and tape by a group of German cleaning ladies in 1973 because they, after a reception of Leverkusen's local Social Democratic Party (SPD) at the contemporary art museum Schloss Morsbroich, mistook the art object for, well, a dirty tub. Beuys received nearly 60,000 Deutsche Mark in compensation and, using photographic images of the initial installation, reconstructed the tub in 1977, including its plaque, announcing that it had served as a bathing vessel for Beuys as an infant. In 2013 the work was gifted to the Lenbachhaus in Munich, yet different stories

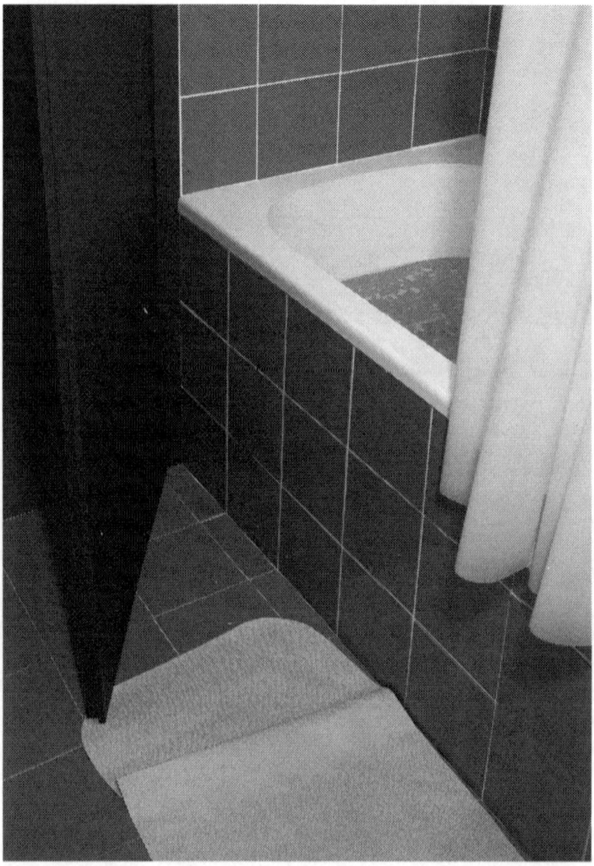

Figure 4.4 Thomas Demand: *Bathroom*, 1997.

about the 1973 incident continue to circulate and poke fun at the place, language, and self-understanding of contemporary art.

Like the interior of cars in the history of film, bathtubs in art have repeatedly served as sites to interrogate grey zones between dominant ideas of the personal and the political, public intimacy and intimate publicness. Tubs often display the subject as vulnerable, withdrawn, and singular, as much as they offer stages to display subjectivity, the body and its surfaces, as objects of willful and unexpected visibility. Nothing in art is safe and innocent about the appearance of tubs. Their very presence seems to provoke the language of seclusion and discretion we tend to dedicate to them in everyday life.

The position of Kjartansson—who is known to play the role of agent provocateur—in the tub thus comes as no surprise: it can easily be seen as an

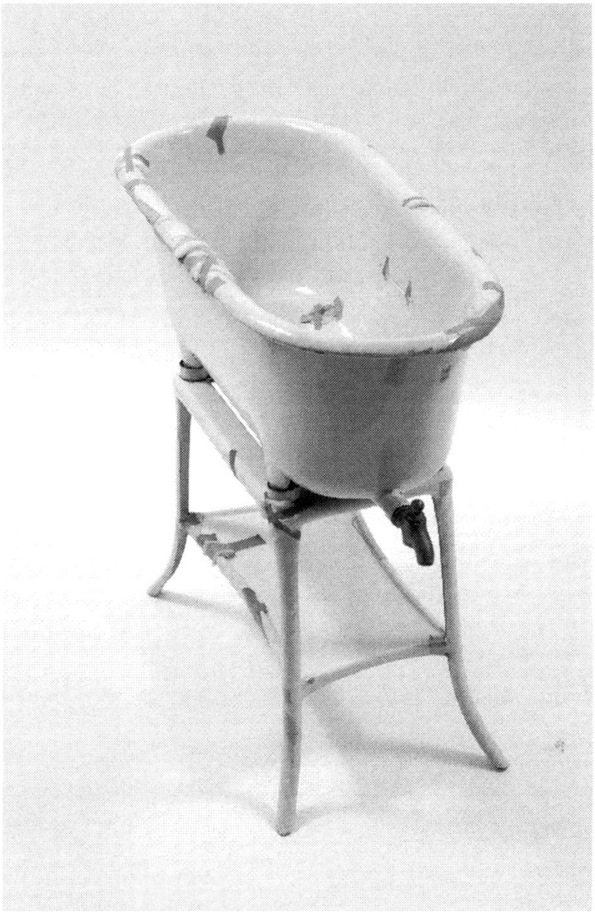

Figure 4.5 Joseph Beuys: *Badewanne*, 1960.

effort to transform the vulnerable into an exhibitionist spectacle of first rank. And yet, the tub—and its water—in *The Visitors* does much more than merely offer a stage for the blurring of boundaries between the intimate and the public. Whether Kjartansson submerges his own body or the body of his guitar into the tub's water, whether he at some point during the performance uses the surface of water as a creative drumming device, or whether he at another point of the loop changes the water level and its temperature—the tub in *The Visitors* offers a laboratory to probe the workings of resonance in both its strictly scientific and its affective or metaphorical understanding. Generously on display, Kjartansson's body and guitar typically attracts the attention of most viewers of the installation. What we cannot afford to ignore, however, is how body and instrument interact with the tub's water—how the use of water in *The Visitors* simultaneously facilitates and allegories the power of vibrational contact, of resonant relationships between animate and inanimate, human and nonhuman matter. In this chapter, I profile three different artistic positions that explore resonances between sound and water before I return once more to *The Visitors* to tease out the significance of Kjartansson's place in the tub, and its particular contribution to the history of bathtubs in art, in further detail.

Water Waves

The pathbreaking work of nineteenth-century physicist Hermann von Helmholtz on sound, music, and hearing paralleled his interest in hydrodynamics, that is, the scientific analysis of the flow of water. In an influential paper published in 1858 entitled "On Integrals of the Hydrodynamic Equations which Express Vortex-Motions," Helmholtz mobilized differential derivations in order to define a framework for better understanding the motion of water—a topic far from abstract and theoretical given the modernizing frenzy of German-speaking lands to reroute rivers and claim usable terrains from aquatic areas in the nineteenth century.[1] What is remarkable about Helmholtz's work circa 1860 is the fact that he understood his interests in multitudinous wave formations of water and in the laws of acoustical perception as deeply connected. In his own words:

> Water in motion, as in cascades or sea waves, has an effect in some respects similar to music. How long and how often can we sit and look at the waves rolling in to shore! Their rhythmic motion, perpetually varied in detail, produces a peculiar feeling of pleasant repose or weariness, and the impression of a mighty orderly life, finely linked together. When the sea is quiet and smooth we can enjoy its coloring for a while, but this gives no such lasting pleasure as the rolling waves. Small undulations, on the other hand, on small surfaces of water, follow one another too rapidly, and disturb rather than please.[2]

In Helmholtz's view, scientists could learn as much from the affinities between water and music as musicians. Water's wave motions helped explain how certain forms of interference, from a scientific point of view, produced musical beats

[1] David Blackbourn, *The Conquest of Nature: Water, Landscape, and the Making of Modern Germany* (New York: W.W. Norton, 2006).

[2] Qtd. in Cahan, *Helmholtz*, 201.

that interrupted the flow of tones and thereby pushed what ordinary listeners appreciated as harmony into a realm of dissonance or disharmony. Likewise, as with the sea's ability to associate distant lands through flows and waves, artists connected to their audiences through the currents of their respective souls, the way in which specific sound waves managed to touch upon the sentiments of their listeners and lifted them to a realm of higher beauty. Water and music, in Helmholtz's perspective, did not simply share and mutually illuminate the laws of wave formations and resonance; their flows and vibrations also represented what defined the underlying spiritual unity of science and art.

No contemporary sound artist comes closer to Helmholtz's interest in the intersections of art and science than Carsten Nicolai. Born in Chemnitz (formerly: Karl-Marx-Stadt), the Berlin-based artist and experimental musician known as Alva Noto has developed numerous installations since the early 2000s

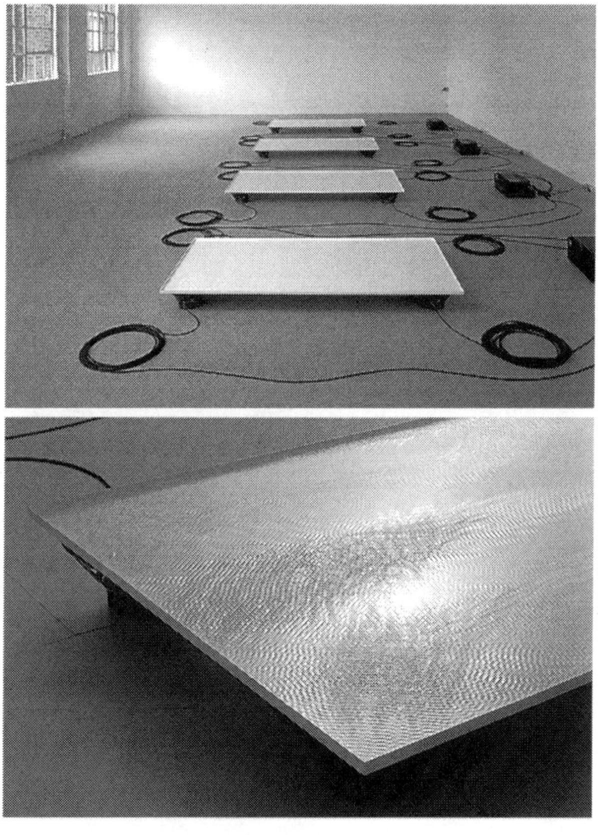

Figure 4.6 Carsten Nicolai: *wellenwanne*, 2001/2003/2008.

in which different wave forms interact with each other—works that probe the audibility and visibility of various light and sound frequencies and whose often minimalist aesthetic brings science, electronic music, and the visual arts into ever-new conversations.[3] First exhibited in 2001, in *wellenwanne* (wave tub) (Figure 4.6), Nicolai explicitly investigated possible couplings between sound and water, the latter used to visualize the former, including shifting patterns of interference caused by different audible and inaudible sound frequencies. The installation's media are four aluminum trays, CD players, CD, amplifier, speakers, and water. Presenting the installation as a model and test arrangement, Nicolai's described the interaction of these different elements as follows:

> Flat trays are filled with water, each resting on four loudspeakers, which transmit the sound compositions via vibrations onto the water surface. The various sound pieces, which are partly inaudible, vary for each tray so that the sound signals generate various changing interference patterns. Based on the unusual physical properties of water this aesthetic-scientific test model touches on areas of particle physics posing the question how sound frequencies, as a form of energy, are able to modulate particles.[4]

At its most basic, *wellenwanne* visualizes processes of resonance: the way in which the vibratory movements of sound, whether its tones are above or below the hearing threshold of the average human ear, incite metal and water to vibrate and in response produce wave motions. The laws of resonance here do not simply translate energy from one sensory register into another, but exemplify the extent to which waves and matter, energy and particle structures can go hand in hand. Resonance converts sound into moving images, the imperceptible into the sensible, the acoustical or visual unconscious into the realm of possible comprehension. In this, *wellenwanne* recalls the eighteenth-century work of Ernst Florens Friedrich Chladni who famously visualized the working of resonant vibrations by drawing a bow over pieces of metal whose surfaces were covered with sand, creating intricate patterns whenever the metal plate reached a state of resonance and thus arranged the sand into nodal patterns, so-called Chladni figures.[5] And yet, what distinguishes Nicolai's work with water from Chladni's

[3] For a useful introduction to Nicolai's work in English, see the 2014 interview with Mark Rappolt on the occasion of Nicolai's opening of a city wide light and sound installation in Hong Kong, *ArtReview Asia* (Spring/Summer 2014), https://artreview.com/features/ara_springsummer_2014_feature_carsten_nicolai/.

[4] http://www.carstennicolai.de/?c=works&w=wellenwanne.

[5] Ernst Florens Friedrich Chladni, *Entdeckungen über die Theorie des Klanges* (Leipzig, 1787). For more on Chladni's contribution to Romantic art and science in Germany, see Myles W. Jackson,

work with sand is a fundamentally different approach to document the work of resonance in and across time. Chladni's resonance figures aspired to transpose the passing of acoustical time into a still image. They not only visualized what exceeded vision, but engaged resonance as a mode of suspending the ordinary workings of linear time. Nicolai's *wellenwanne,* by contrast, approaches time as the very medium in which resonant transpositions and vibratory exchanges unfold. Its point is not to arrest flow, but on the contrary feature flow as both a condition and an effect of the process by which different particles touch upon and animate each other.

Things are even more complex, though. The original installation of *wellenwanne* featured four speakers attached to each corner of the aluminum tub, each triggering intricate ripple patterns in the water in response to their low-frequency drone. As different frequencies cause different intensities of vibration, we not only observe the dissemination of different ripple patterns from each of the tub's corners, but unpredictable formations of wave interference in the middle of the tub. As a visible analogue to the invisible work of sound, the medium of water here serves as a host to invite some interactions we may want to deem congruent and harmonious and others that may strike the eye as unruly and dissonant. The math of resonance is far from simple: 1+1+1+1 does not necessarily add up to 4; the interfering energies of waves and particles may in fact cancel each other out and result in 0, that is, waveless and seemingly unmoving water surfaces, or produce tumultuous movement that seems to defy measurability altogether.

A 2012 version of the installation, entitled *wellenwanne lfo* (Figure 4.7) and shown at MoMA's aforementioned *Soundings: A Contemporary Score* in August 2013, puts these propositions to a further test. Again, here's how Nicolai himself describes the work:

> A two-channel composition of sub frequency sounds is transmitted onto the surface of a specially designed water pond via four exciters through variations in air pressure. Each sound continuously creates concentric circle waves causing interference patterns when they meet. By means of synchronizing the sound waves with a stroboscope the wave patterns can be made visible on a display screen. Depending on the original frequencies induced on the water surface the interference patterns can be either regular or irregular hence creating orderly or rather chaotic visual results that cannot be entirely controlled.[6]

Harmonious Triads: Physicists, Musicians, and Instrument Makers in Nineteenth-Century Germany (Cambridge, MA: MIT Press, 2006).

6 http://www.carstennicolai.de/?c=works&w=wellenwanne_lfo.

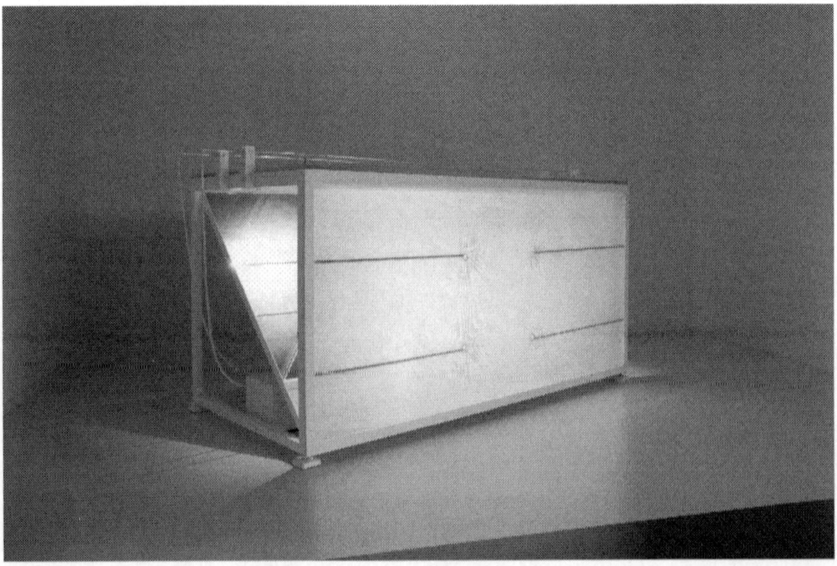

Figure 4.7 Carsten Nicolai: *wellenwanne lfo*, 2012.

Throwing stroboscopic light impulses into the mix, *wellenwanne lfo* vastly complicates the work of resonance and interference staged in the original *wellenwanne*. Far from simply picturing the invisible waves that make up sound, *wellenwanne lfo*—similar to the first version—aspires to embody their vibratory energy, yet it does so while at the same time blurring presumed differences between the audible and the visual, sound and image. What we see and hear is no mere process of translation, of building bridges and precisely thereby reifying categorical distinctions. The installation instead maps different logics of refraction and reflection onto each other in such a way that sound becomes image, and moving images assume the qualities of sound. In *wellenwanne lfo*, sound and image are designed to resonate with each other, similar to how quantum mechanics considers the physics of waves and the physics of particle matter as two sides of the same coin. This meshing of ontological and phenomenological registers invites the audience to approach Nicolai's upright water tub either as an object of scientific curiosity or as a resource of meditative contemplation. Or, in fact, as both at once: as a laboratory in which the aesthetic and the cognitive interfere with, yet also complement and resound with each other.

Recalling the work of Andrei Tarkovsky, Michel Chion writes: "In his use of sounds of dripping water . . . Tarkovsky never stopped looking for that point of semimusicality where the cosmos or nature seems on the brink of speaking or singing, without us ever being sure of this."[7] In Nicolai's work with water, the cosmos remains largely silent due to the artist's use of mostly subsonic sounds. Though typically exhibited under the label of sound art, his installations appear to cater at least as much to his audience's eyes as to their ears. And yet, water—its flows and ripples, its undulating waves and vibrating matter, its resonant qualities and rhythmic motions—here presents an analogue of music: a medium energized by sound, a silent medium at the threshold of bursting back into song and music. Nicolai's water is deeply ambivalent. It can be seen as wave or particle, static or in motion, representational or presentational. It lends a body to the invisible, reveals what typically remains concealed. At its core, however, Nicolai's tubs showcase water as a unique medium of extraordinary hospitality, even if—or precisely because—no human agent is directly involved in the final execution of their interference patterns. Whether upright or horizontal, these tubs provide spaces in which different materialities, bodies, rhythms, and movements meet, touch upon, coexist or interact with and, in so doing, shed light onto each other. They reveal the extent to which science shares some common ground with art and the extent to which aesthetic experience need not be averse to scientific inquiry. Just as Nicolai's different water and wave tubs allow sounds and images to resonate with each other, so his installations find the unity of physics and metaphysics, of scientific laws and aesthetic affects, not in what Helmholtz considered the human soul, but in the vibrational dynamics that resound in and animate all matter.

Interspecies Resonance

Nicolai's water tubs enable resonant interactions between sound and light well aware of and deliberately drawing on the fact that sound and light waves are of a very different kind. Light waves, physics tells us, oscillate in parallel to the direction they move. They are transversal in nature, meaning that they progress in space like ocean waves approach the beach. Light waves travel at extremely high speed, slow down when entering liquids of higher density, and of course do not penetrate solid matter at all. Sound waves, by contrast, vibrate along the direction of their movement. They are longitudinal in nature, that is, they move

[7] Michel Chion, *Sound: An Acoulogical Treatise*, trans. James A. Steintrager (Durham: Duke University Press, 2016), Kindle file.

forward by compressing and expanding molecules along their path. Think of their movement and shape analogous to what people in gyms do when working out with ropes. While the speed of sound waves in the air and open space is considerably slower than that of light, the velocity of sound accelerates the denser the medium of travel (even though it may lose some intensity in the process).

In Nicolai's work, water serves as a meeting ground between light and sound, a resonant medium hosting the encounter of different logics of vibratory movement. The specificity of water as a medium of sonic transmission and acoustical perception is at the core of the next project I profile in this chapter as well. Unlike Nicolai, however, our next artist—musician David Rothenberg—is no longer concerned with the differentials of sound and image. His focus is first and foremost on the aural properties of liquids. Moreover, in his efforts to jam with humpback whales, Rothenberg moves beyond the relative absence of non-mechanical agents in the performance of Nicolai's works as he radically expands the medium of experimentation: from the limited cubic footage of Nicolai's tubs to the oceans of the Earth, defining almost three-quarters of our distressed planet's surface.

"Well into the nineteenth century," writes Jochen Hörisch, "when one spoke of media, one typically meant the natural elements such as water and earth, fire and air."[8] To which media philosopher John Durham Peters adds in an effort both to understand media as environments enabling the possibilities of being and to think of the sea as a very unique infrastructure of agency and perception:

> The sea is a habitat as varied as earth, air, and sky. It has subfreezing polar waters and boiling temperatures near heat vents, oxygen-rich zones and dead spots, translucent surface waters and lightless depths. It has sustained vast epochs of evolutionary experimentation. If media theory concerns the different sense ratios through which mind interacts with world and the various worlds that come into being in distinct historical and ecological climates, the ocean should be of primary interest as an environment that invites us landlubbing bipedals to abandon most everything we take for granted. The marine world invites fundamental anthropological questions.[9]

Rothenberg's approach to the underwater songs of humpback whales wants to ask such questions. Its desire is to establish resonant interactions across species lines so as to hit upon melodies and rhythms that are neither human

[8] Jochen Hörisch, *Ende der Vorstellung: Die Poesie der Medien* (Frankfurt: Suhrkamp, 1999), 134.
[9] Peters, *The Marvelous Clouds*, Kindle file.

nor cetacean in nature, but—like what Peters understands as media in times of the Anthropocene—hover somewhere in between culture and nature, the earlier seen as that which may carry and communicate meaning, the latter recognized as the background, the material condition of the possibility, of whatever might count as meaning at certain historical junctures. Though informed by the science of underwater sound research and marine zoology, Rothenberg's principal aim is not to establish a formal language of interspecies communication or to provide new insights about why, how, and to what end humpbacks sing. Instead, his work considers the prospect of musical resonance across species lines as an experiment in aesthetic knowledge production, relying on what—as Kant would say—the purposiveness of art's purposeless objects and practices as much as the development of communications that evade the codes and arbitrary registers of what we call language.

The astonishing singing of humpback whales was not really recognized by humans as such until the 1960s. Once discovered, however, it came to play a critical role in global efforts to protect whales from extinction. Though scientists have reached no agreement about why humpbacks sing, what is known is the fact that humpback songs are of incredible complexity and duration, and much more musical in structure than the navigational clicking of dolphins or other whales. Examined from a musicological perspective, these songs

> consist of repeating patterns, hierarchically organized at the level of unit (or motif), phrase, theme and song. Each complete song consists of five to seven themes. Some of the phrases end with the same contrasting sound, so they can be said to rhyme, in a way analogous to human poems. A series of these songs can be repeated extensively, up to 23 hours in a single session.[10]

What is perhaps most remarkable about humpback singing is the fact that their song, due to the material conditions of aquatic sound transmission, can at times be heard for hundreds of miles across the ocean, that humpbacks are known to adapt motifs and phrases during a typical season, and that they do so collectively across vast stretches of the oceans, as if one listened carefully to another and integrated the patterns of their cries, groans, wails, gulps, and vibrating upsweeps into their own singing.

Enter Rothenberg, equipped with a clarinet, a microphone, headphones, underwater speakers, a high-quality hydrophone, and a sound recorder (Figure 4.8).

[10] David Rothenberg, "Whale Music: Anatomy of an Interspecies Duet," *Leonardo Music Journal*, 18 (2008): 47.

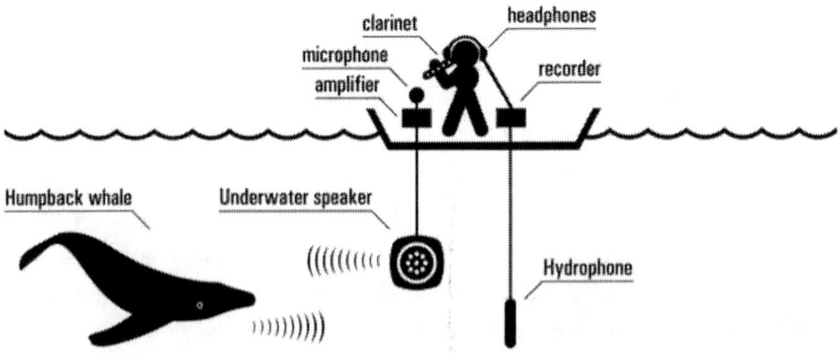

Figure 4.8 David Rothenberg: "How to Play Clarinet along with a Humpback Whale," 2008.

The first two months of 2007 Rothenberg spends off Maui, Hawaii, setting up what it takes to make music across species line and use the particular sonic qualities of water to conjure sounds neither whales nor humans have produced or heard as such. Sonogram devices travel along to chart possible interactions for future analysis. Humpback sounds have tremendous range in frequency, from about 100Hz at the lower end (right above the threshold of average human auditory perception) to about 5,000Hz, from deep and clear sounds that carry virtually no overtones to high warbles and shrieks whose intensity if listened to in close aquatic proximity might easily feel like an assault on a human listener's ear. According to Rothenberg, most of his attempts to get through to the whales don't really go anywhere, with his clarinet doing its thing, the humpbacks near the boat doing their thing. Rothenberg plays mostly high, long-held notes on his clarinet, with sonic propagation underwater producing richer textures of overtones than would be audible on land. During one session, his clarinet resounds shriekier and more uneven, closer to what Rothenberg himself considers the song of whales. At which point, he suggests with some hesitation, audio recording and sonogram document a possible response, a brief moment of resonant interaction:

> What does the whale do? Does this sound become more clarinet-like during the encounter? I am not really sure, but some of our high squeaks are quite hard to tell apart. The clearest sign of communication comes when I stop; he resumes with a direct sense of response, in some cases continuing the very same note I just finished. In other cases he tries to join in and overlap me with a complementary sound.[11]

[11] Rothenberg, "Whale Music," 50.

What does the whale do? Rothenberg is not sure both during the jamming session and when later carefully studying the recording and the printed graph of his sonograms. What is sure, however, is that this effort to attune the sounds of humpbacks and clarinets represents no mere effort of a human to shed the burdens of civilization and "go primitive." On the contrary, at each juncture of the project different processes of mediation reign central and are being deliberately put into play: to jam with whales is to reckon with and rely on how media such as water, audio transmitters, hydrophones, recording technologies, and sonograms provide uniquely imbricated environments of perception and communication. Elemental or technological in nature, media here do not simply define channels of possible exchange. Instead they offer alternative infrastructures of audition and perception and modify the sense ratios through which minds perceive the world. To jam with humpbacks, in Rothenberg's understanding, is to partially attune the human sensorium to conditions in which animated beings neither listen with ears nor speak with tongues, yet seem to know how to see with sound, hear with bones, and connect across hundreds of miles. The marine environment, Peters argues, "is a superb place for sound studies."[12] To play music with whales is to access a vibrant media ecology in which sound operates quite differently from what we think it to be and do above the sea's surface. Whether the sounds of Rothenberg and the humpback really meet each other or not, what in the end matters about his project is the extent to which it embraces different types of media, including the one of water, to explore and brush against the biological and cultural limits of human sound production and perception; and how this project, in so doing, asks tough anthropological questions about what humans typically take for granted about the modalities of sound production and reception. The whale's song causes Rothenberg not only to access the world through unknown sounds. It inspires sounds that reshape what we call and understand as world in the first place.[13]

[12] Peters, *The Marvelous Clouds,* Kindle file.

[13] Here's how Rothenberg describes what happens when human and cetacean voices entangle with and attach to each another:

> I'm screaming, wailing, screeching too, trying to pack as much emotion as I can into one moment when a whale wants to listen. It roars through the seas, only for a few minutes, this music neither human nor humpback could create alone. Yet each musician makes space for the other, our duet an overlap of themes from different worlds, human and cetacean. Undersea and above, it is music with no beginning or end. We pause only to back away from the possible sounds that still remain. Then we stop paying attention to one another, back in our separate worlds. Neither human nor whale has forgotten the song we made together (David Rothenberg, *Thousand Mile Song: Whale Music in a Sea of Sound* [New York: Basic Books, 2008], 239).

Water against Violence

As something whose immateriality is hard to capture and hold on to, and therefore poses unique challenges to be studied, sound—in Brandon LaBelle's words—should be considered a weak object: "I search for it, and yet, it is already gone; even if recorded, I must play this sound again and again in order to understand its shape and density, its frequencies as well as psychoacoustical impact. As such, it may slip through the fingers to elude description."[14] Sound's weakness is a weakness of the listener's memory, our attention's structural inability to tend to and retain passing figures in time over extended durations. While such material weakness may at the same time constitute the strength of music, which is its capacity to elicit strong affects and attachments, the history of music making has witnessed numerous techniques to strengthen sound's structural weakness and endow it with greater density and shape and, hence, to protect it from slipping through the listener's finger.

One such technique is the use of rhythm. Or to be more precise: the effort to play out the power of rhythm against the temporal scaffolding of musical meter. What we call musical meter offers a structure of temporal entrainment that allows musicians as much as listeners to develop and synchronize their perceptions of rhythmic figures, that is, the beat, feel, patterning, and groove of temporal manipulations that drive particular melodies or accompaniments and press against the metrics of measurable time.[15] At once enabled by and challenging the regime of countable time, musical rhythm on the other hand produces patterns of perception and cognition, of memory and anticipation, that counteract the presumable weakness of the musical object. As it plays with and refuses the dictates of meter, rhythm sets in motion dynamic energies that open up the listener's limited window of temporal attention—their ability to tending to the repetitions and variances of a given musical piece. According to Mari Riess Jones: "We continually cast ourselves forward by rhythmically anticipating future events that may occur within small and larger time intervals.

We may never know what whales sense or do when singing their songs, whether their voice resounds and modulates the human's sound or not, whether or not they adapt to the sounds of the sea amid their own sea of sounds. What we do know from Rothenberg's practice, however, is that the event of interspecies resonance—whether real or imagined—has the potential to retune land-based sensoria and thereby decenter visions of the world that solely pivot around the human. Whether whales really sing with humans or not, their sea of sounds has the power to remind us of our own hubris and limitations. It reveals the self-destructive conceits of subjectivity in an age in which human civilization has come to define the entire geological and marine conditions of life on this planet.

[14] LaBelle, *Sonic Agency*, 20.

[15] For a comprehensive study of musical meter and temporal perception, Justin London, *Hearing in Time: Psychological Aspects of Musical Meter* (Oxford: Oxford University Press, 2004).

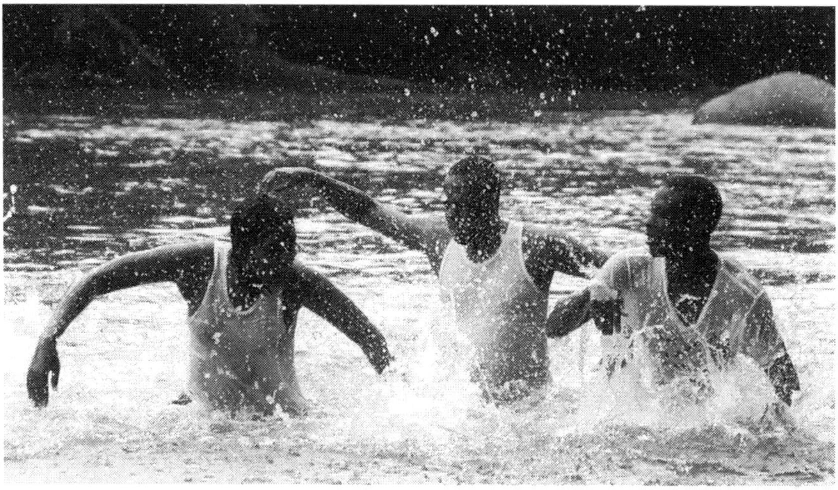

Figure 4.9 Marcos Ávila Forero: *Atrato*, 2014.

These paths form the patterns of mental space and time and so can establish for us that sense of continuity and connection that accompanies comprehension."[16] Rhythm lends a crutch to memory and attention that meter itself cannot provide. It enables musicians and listeners to explore music's weakness as its strength and attach body and mind to what eludes solid shape. As a patterned movement of sound, rhythm thus allows musical time to resonate with larger temporalities and histories, with their respective calibration of chronological and qualitive dimensions of time as much as their particular infrastructures of the linear and the circular, of experiences of progress, repetition, continuity, and rupture.

Named after a river that cuts through Colombian territories ravaged by armed hostilities and ongoing violence, Marcos Ávila Forero's 2014 research and performance project *Atrato* (Figure 4.9) uses the waters of the Río Atrato as a rhythmic instrument drumming up against the traumatic temporalities of the present. The Atrato once served as a main route of trade and migration. In recent times, however, it has transformed into one of the main arteries of brutal conflict, a body of water in whose vicinity violent deaths have become the order of the day.[17] Supported by a team of anthropologists, musicians, and

[16] Mari Riess Jones, "Only Time Can Tell: On the Topology of Mental Space and Time," *Critical Inquiry*, 7 (1981): 571.

[17] For more on historical transformation of the Río Atrato into a political and ecological minefield, see Frederick Gillingham, "Toxic River: Mining, Mercury and Murder Continue to Plague Colombia's

ethnomusicologists, French-Colombian artist Forero asked a group of Afro-Colombian residents from the region to revive old traditions in which local residents jointly hit the river's surface with their hands and arms to produce complex rhythmic figures able to reverberate over long distances. As they seek to synch their drumming, their rhythmic work at once mimics the sounds of guns and explosions, but in mimicking and absorbing the sounds of everyday violence the drumming at the same time seeks to contain this violence, move with it beyond it.

In a video of the project shown at the Venice Biennale in 2017, we see different groups of local residents standing up to their hips and waists in the river's water and hitting its surface. It quickly becomes clear that the effort of synchronizing one's beat and resonating with each other is as challenging a feat as the sheer physicality of the exercise. While we see kids playfully swim or rafts pass by in the background, the drummers thrash the water in the foreground, widely swinging their arms to gain momentum during the impact, carefully listening and glancing sideways to attune their own sound production to that of their neighbors. A group of seven is shown during an instructional session. Some comment on the work it takes to find the desired rhythm's groove. "I got lost! The bass went!" "There, we are back!" "No, we left again . . . and we arrived," and "No, no he hasn't achieved it" we hear individuals shout before what appears to be their instructor reminding them: "We're going to keep the bass, you need to have the rhythm pattern very clear. Give it bass, go on!" Another group, first involving two then three individuals are shown in another section of the river, exploring different rhythmic patterns without the help of a visible instructor. They admit to each other the strenuous nature of what for others might look like children's play. Eventually, some figures approach our group from behind and declare "Hey! We are going to join! Let's join!" and add rhythmic complexity to what the three have established already.

The setting of Forero's drumming amid the Chocó forest evokes a seemingly timeless cliché describing time and life as a river. We often use this metaphor with good intentions: to express the fluidity, historicity, and transformative nature of all matter. But in truth, the image suggests a certain automatism of change, a uniform and predetermined progress of something from its origin to end.

Atrato," *Mongabay* (October 2019), https://news.mongabay.com/2019/10/toxic-river-mining-mercury-and-murder-continue-to-plague-colombias-atrato/; and Allan Gillies, "Humanitarian Crisis, Dignity and Hope on the Río Atrato," *The Drouth*, http://www.thedrouth.org/humanitarian-crisis-dignity-and-hope-on-the-rio-atrato-allan-gillies/.

Things might constantly undergo change, but change itself is preprogrammed, no alternative paths can be taken, no real surprises and sudden inversions will happen. According to the image of time as a river, fluidity and history are fate.

As they reanimate traditional musical patterns, the water drummers of Forero's *Atrato* disturb this image of water as destiny and fate. Their drumming unhinges the cliché of history as inevitable, embracing sound and water as media to transform the traumatic landscapes of their present into unruly dominions of time. The drummers' drumming does not simply seek to resound past patterns of music making so as to generate a present sense of community; it wants to echo and by echoing to break the spell contemporary sounds of bloodshed place on the lives of local communities. Forero's ambition is to rechannel seemingly unchangeable burdens of history and train our attention toward what may transcend the experience of the everyday as a perpetual metronome of violence. Rhythm here helps attend to the promise of forgotten pasts as much as its resonant vibrations hope to clear the ground for the possibility of a better future. A future that emancipates the present from the burdens of fate. The only kind of future that deserves to be called future at all.

Ragnar's Bathtub II

In Kjartansson's native Iceland, tubs and pools are much more than mere sites of personal hygiene. Often heated by thermal energy, they are part of what one might want to consider an alternative public sphere, equivalent to the eighteenth-century salon in Britain, France, and Germany or the piazza in Renaissance Italy (Figure 4.10).[18] Iceland's public pools are known as places to mix and mingle, catch up with your neighbors, connect across social divisions, and discuss politics and public affairs, with their warmth allowing outdoor exchanges Iceland's harsh climate tends to prohibit otherwise. It's where you go, not simply to wash off at the end of a working day, but to chat and relate, to debate and mull over what exceeds the contours of one's individual life. Pools and smaller tubs in Iceland are private places with public relevance. It therefore should come as no real surprise that Kjartansson chose Rokeby's tub to intone *The Visitors*' 64-minute lament and assume his position within the network

[18] Dan Kois, "Iceland's Water Cure: Can the Secret to the Country's Happiness be Found in Its Communal Pools?" *The New York Times Magazine*, April 19, 2016, https://www.nytimes.com/2016/04/24/magazine/icelands-water-cure.html.

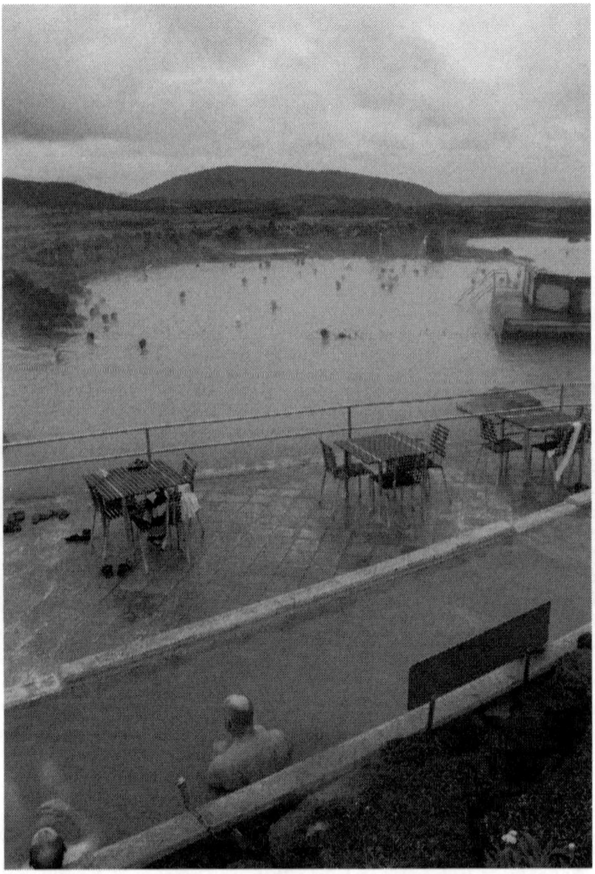

Figure 4.10 Thermal Pool in Iceland (Jarðböðin við Mývatn).

of the work's musical nodes. Though it is easy to think of his place in the tub as a stage for bohemian exhibitionism, we do much better to consider it as an interface: the point at which different orders or logics of life—the private and the public—interact with each other and through their interaction negotiate their difference as much as their mutual dependencies.

And yet, Kjartansson's tub is much more than just a site, stage, or interface hosting encounters with and touching upon his audiences. Throughout the performance, it also serves as a musical instrument in its own right, whether its water modulates the sounds of Kjartansson's submerged guitar and body or is being used as a drum set itself. Think therefore of this tub as a resonance chamber amending and complicating the acoustical vibrations Kjartansson seeks to feed into the microphone installed next to him and thereby relate to the other members of his band. And as such, it resonates with what contemporary sound artists such as Carsten Nicolai, David Rothenberg, and Marcos Ávila Forero do when exploring and putting to work the sonic qualities of liquids or when approaching the elemental medium of water as an analogue to the medium of sound. To be sure, no hidden bass speakers here make Kjartansson's water vibrate, no hydrophones pick up strange sounds, no rhythmic thrashing absorbs and reworks the noise of guns. But Nicolai's interest in wave interferences, Rothenberg's hope to decenter the place of human-based sound production, and Forero's ambition to animate memory and mobilize attention against the rhythms of an unbearable present—all this stirs Kjartansson's water in one way or another as well and defines his tub as a laboratory of acoustical experimentation, discovery, and intervention.

Water is sound art's resonant other. It is its secret and not so secret ally. It can be used to visualize the physics of sound propagation, but it may also bring into focus sound art's metaphysical aspirations, that is to say, the many ways in which sound in art today features heard and unheard vibrations as media to transform the inhospitable environments humans have come to engineer for their and their planet's life.

5

Air 2.0

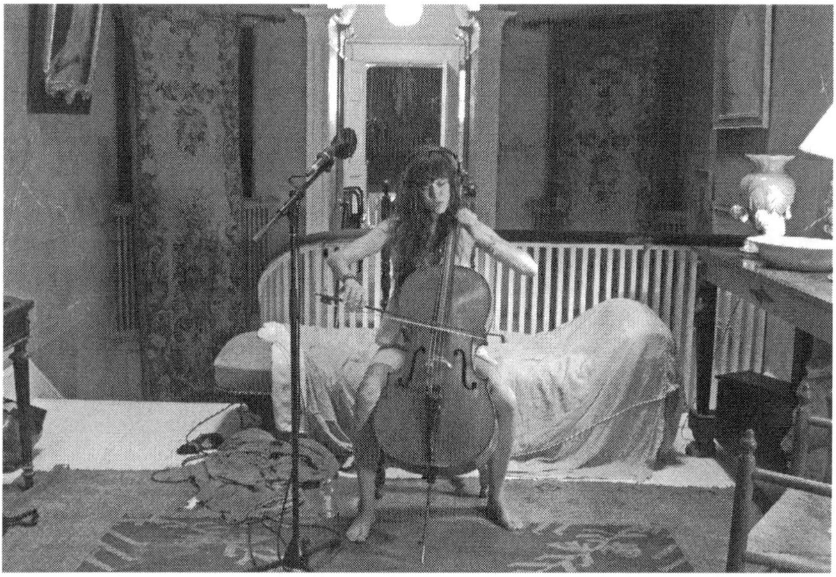

Figure 5.1 Ragnar Kjartansson: *The Visitors*, 2012 (Gyða Valtýsdóttir). See also Plate 4.

Baroque Redux

Most of what is commonly said about music takes the propagation of sound through air for granted, and it is air's presumed immateriality that often energizes our belief that we can control, grasp, and in fact possess musical sounds. To approach sound through the medium of water, as we did in the previous chapter, unsettles such presumptions. It urges us to rethink music as vibrational practice, and in doing so it resituates the role of human agents—in Nina Sun Eidsheim's words—"from conqueror of the figure of sound, to plain member and transmitter of a vibrational field."[1] The effort to consider sound as a resonant relay of different materialities redefines music as a realm in which everything assumes the status of a medium and nothing exists free of mediation. Though it may at first simply represent an epistemological intervention, this rethinking at the same time unlocks sound as a field of ethics and political ontology. It reminds us of the fact that sound can neither be fixed nor owned by isolated individuals; that music does not exist independent from the materiality of its transmission; that the effects of sound propagation are never stable nor entirely knowable; and that we therefore may want to consider music makers as stewards rather than sovereign originators of sonic material—as nodes within larger networks of mediation that simultaneously rely on and impact other nodes without ever assuming full control over given processes of vibrational transaction.

Our attention to Kjartansson drumming his bathwater, then, is no mere detour in trying to understand why and how *The Visitors* moves its audiences. It goes to the heart of the matter—of why and how the matter of sound matters. Kjartansson's drumming urges us to think hard about how the work's sounds spread, resound, reverberate, and enable different bodies—animate and inanimate—to touch upon each other; and it directs our curiosity to the different layers of mediation that make the performance possible as a whole, including the seemingly neutral medium of air and the kind of spaces it fills, connects, and separates. To approach music as vibrational practice urges us to pay heed to how sound in *The Visitors* inhabits air and other media to define ever-changing spatial relationships. It asks us to illuminate the ways in which sound travels between the domains of the seen and the unseen and thereby engages dreams, desires, and fantasies of music as something that plays on us

[1] Nina Sun Eidsheim, *Sensing Sound: Singing and Listening as Vibrational Practice* (Durham: Duke University Press, 2015), Kindle file.

and plays itself rather than as something that is being played and produced by human agents. Like Kjartansson's water, the band's microphones and earphones play a critical role in showcasing the extent to which sound here cuts across multiple media and sensory registers. In spite of their apparent technological functionalism, they—as both prostheses and allegories of air—resonate with various metaphysical subtleties and wondrous enigmas.

Kjartansson, of course, is not the first to stress the fact that nothing in music precedes the materiality of mediation. He is not the first either to play with the idea of invisible and/or self-producing sound. One of the first and influential precedents of this can be found in the work of seventeenth-century polymath Athanasius Kircher whose 1673 treatise, *Phonurgia Nova,* discussed various technologies designed to disseminate sound in often astonishing ways. A true representative of the Baroque, Kircher's work hovered along the border between magic and science, between the marvelous and the coldly calculated. Few of Kircher's sound media of the mid-1600s have received more attention than his aeolian harps, intended not only to astonish listeners with unheard-of sounds, but to emancipate the audible from human intentionality and precisely thus advance the analysis of sound production and mediation (Figure 5.2). Kircher envisioned wind harps as media whose exact operations, like those of many of

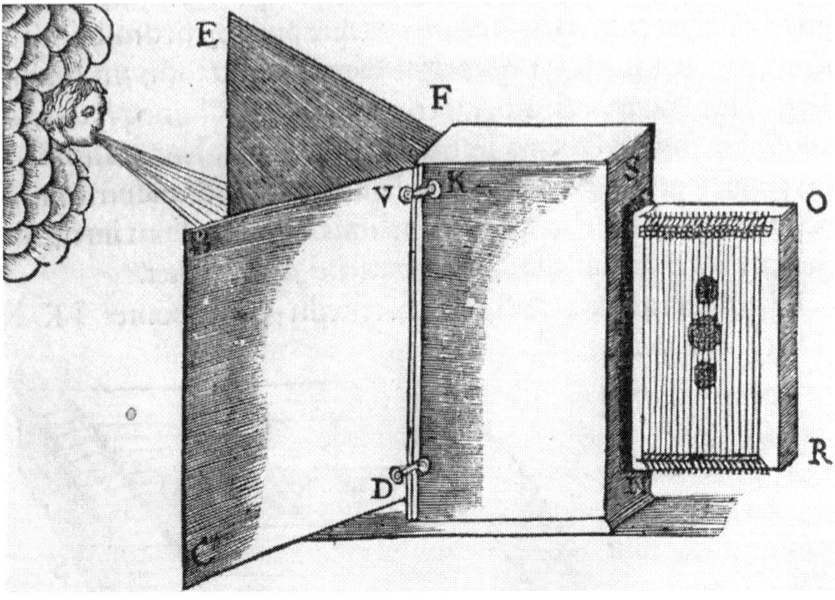

Figure 5.2 Athanasius Kircher, *Phonurgia nova*. Campidonæ, 1673.

his other audio machines, were hidden from the listener's sight. Their invisible sounds engaged what twenty-century scholars have come to call the acousmatic, a severing of visual and acoustical perception able to galvanize the perceiving subject's attention, imagination, and curiosity.[2] And their legacy, as we will see in this chapter, is helpful to think through the aesthetics of both mediation and (in)visibility in *The Visitors*.

In the work of nineteenth-century romantic poets such as Goethe, Eichendorff, Shelley, and Mörike, aeolian harps played a prominent role to bridge perceived gaps between nature and civilization, the divine and the human. Their poems frequently featured forms of music produced by the forces of the wind, driven by the desire to develop allegories of the poetic in which nature could appear as art and art as nature and precisely thus achieve their fullest and quasi-sacred unfolding. Kircher's much earlier experiments, on the other hand, were not yet concerned with tuning into God's inaudible language to consecrate the workings of human creativity. His focus was more on astonishing unsuspecting listeners with acoustical experiences for which no category existed. While the harp conjured, in Kircher's own words, pleasant tones, their foreignness was meant to activate the listener's sense of wonder. The harp's affective echoes—what Kircher called "Bestürzung" (consternation, perplexity)—at the same time inspired Kircher's own systematic analysis of acoustical phenomena, Kircher's pathbreaking preoccupation with echoes, reverberations, sound production, transmission, and reception. Rather than soliciting speechless reverence to the divine, Kircher's audio media provided the emotional resources to explore how sound moves and how it moves the listening subject. Instead of dumbfounding the listener, Kircher's "Bestürzung" set minds into motion.

There were no aeolian harps present when Kjartansson & Co. recorded *The Visitors* at Rokeby Farm, no forces of wind wondrously activating gadgets designed to remain invisible to later audiences in the gallery setting. The strings, piano and accordion keys, drums, and cannons in *The Visitors* instead are all, visibly and seemingly deliberately, being manipulated by skillful human hands, each individual screen singling out the agency of each musician involved. As they play their cellos and guitars, the somewhat extraterrestrial figures of Gyða and Kristin Anna Valtýsdóttir may appeal to romantic fantasies of music using musicians as uncanny mediums. But both are clearly also shown as being on top

[2] Pierre Schaeffer, *Traité des objets musicaux: essai interdisciplines* (Paris: Éditions du Seuil, 1966); Michel Chion, *Audio-Vision: Sound on Screen*, ed. and trans. Claudia Gorbman (New York: Columbia University Press, 1994).

of their musical proficiency, as highly competent performers exactly knowing what they are doing. As he fires a century old cannon on the lawn of Rokeby Farm, Richard Aldrich may employ an object typically not considered a musical instrument. But nothing about his production of sound comes as a marvelous surprise as we can witness each step of him preparing the firing in great detail on one of the nine screens. Nothing, in *The Visitors*, we may want to conclude holds the power to perplex and consternate the listener. No acoustical media here hide their operation in such a way that audiences may consider sound as self-producing, as a marvelous reality sui generis.

And yet, Kircher's fascination with the wonder of tones traversing the medium of air and the lines between the seen and the unseen nevertheless has found intriguing echoes in the performance of *Feminine Ways*. *The Visitors*' unseen strangely resides in plain sight. It helps structure the audible without playing tricks on our senses. And what produces wonder here is how acoustic media—the network of wires, plugs, circuits, membranes, diaphragms, and amplifiers that connect the mansion's different microphones, speakers, and rooms—manage to allow musicians to intone sounds *as if* these sounds were mere echoes of remote forces and events; *as if* no sound could exist without resounding distant sounds; *as if* the best way to make music was to allow vibrational energies to play on and resonate through you and your instruments; *as if* performers, at their best, should think of themselves as nothing but media chiming into the tunes of unseen others. What produces wonder here is how eight musicians understand how to build a common house of sound and air in spite (or because of?) the fact that all of them dwell in places of physical separation, cannot see each other, and are—unlike their viewers—tied to the acousmatic.

It is easy to deride such "as ifs" as mere flirts with the phantasmagorical. Things are more complicated, however. For *The Visitors*' point is not to reenchant a modern age that has proudly disavowed any remaining belief in the wondrous, but to reveal the extent to which moments of enchantment remain integral to and deeply embedded in contemporary life. "To be enchanted," Jane Bennett argues, "is to be struck and shaken by the extraordinary that lives amid the familiar and the everyday."[3] It means to be charmed and disturbed at the same time, to be fascinated by what exceeds anticipation in the midst of the ordinary, and to be mesmerized by what even in face of the most mundane and repetitive destabilizes our habitual sense of order. *The Visitors*' play with aeolian fantasies

[3] Bennett, *The Enchantment of Modern Life*, 4.

and unseen sounds features the seemingly banal propagation of sound in air and space as music's perhaps greatest source of wonder and enchantment, as a well of what Kircher once called *Bestürzung* and whose literal translation may simply read: "to find oneself falling." Though little seems to happen throughout the entire length of the performance, the work constructs resonant architectures of sound in space and air whose blending of the cultural and the natural, of electronic and elemental media, reveals the continued possibility of experiences of enchantment amid the very strictures of contemporary existence.

Aeolian Variations

Romantic poets embraced aeolian harps because they, as stringed instruments, provided powerful models of the human nerves, the operations of sensory perception, and their respective impacts on the production of poetry. Wind harps offered intriguing examples of the transduction of energy, of how to create harmony between different entities and infuse the human-made with the vibrations of the natural. Though much of the writing on aeolian harps in the nineteenth century approached them as media of timeless meanings and forces, their popularity reflected urgent needs to address the impact of modern technologies on the functions of the human mind and sensory registers, on what it meant to be human in face of rapidly changing technological developments.[4]

This is not the place to discuss the rich iconicity of wind harps in Romantic poetry, their role in encoding art as nature and nature as art. What deserves

[4] As Trower summarizes:

> Powerful, overwhelming stimuli from the external world, such as the wind, could be experienced as both painful and pleasurable, and seemed to have both detrimental and beneficial effects on bodily and mental health. The sublime force of the wind could painfully excite the mind out of dejection and inspire poetic production (within the limits beyond which it could be harmful), and further, it could become a life-giving energy with the power to reanimate life and consciousness. The Romantic mind—with its inspired thoughts, productive imagination, passionate feelings—is not conceived here as spiritually transcending the material body and world. Rather, the harp became one of the chief emblems of a culture in which new technologies helped to produce and served to demonstrate a mechanized understanding of mind in relation to the material world.

Shelley Trower, *Senses of Vibration: A History of the Pleasure and Pain of Sound* (New York: Continuum, 2012), 36. For more on the figure of Aeolian harps in British romanticism, see also Thomas Hankins and Robert Silverman, "The Aeolian Harp and the Romantic Quest of Nature," in *Instruments and the Imagination* (Princeton: Princeton University Press, 1995), 86–112; and Timothy Morton, "Of Matter and Meter: Environmental Form in Coleridge's 'Effusion 35' and 'The Eolian Harp,'" *Literature Compass*, 5.2 (2008): 310–35.

mentioning, however, is how nature philosopher David Henry Thoreau, when coming across the sounds produced by vibrating telegraph cables in the early 1850s, recalled the model of the aeolian harp to describe the networks of industrial modernity, not as the downfall of nature and the aesthetic but, on the contrary, as forcefields redefining what we should understand as nature in the first place. In a diary entry of 1851, Thoreau noted:

> Yesterday and today the stronger winds of autumn have begun to blow, and the telegraph harp has sounded loudly. I heard it especially in the Deep Cut this afternoon, the tone varying with the tension of different parts of the wire. The sound proceeds from near the posts, where the vibration is apparently more rapid. I put my ear to one of the posts, and it seemed to me as if every pore of the wood was filled with music, labored with the strain, as if every fiber was affected and being seasoned or timed, rearranged according to a new and more harmonious law. Every swell and change or inflection of tone pervaded and seemed to proceed from the wood, the divine tree or wood, as if its very substance was transmuted. What a recipe for preserving wood, perchance, to keep it from rotting, to fill its pores with music! How this wild tree from the forest, stripped of its bark and set up here, rejoices to transmit this music! When no music proceeds from the wire, on applying my ear I hear the hum within the entrails of the wood, the oracular tree acquiring, accumulation, the prophetic fury.[5]

Thoreau's ruminations of 1851 invite additional reflection. It is quite impossible to distinguish between what Thoreau, when listening to the wind in the woods, considers as medium and what he considers as subject of mediation. What is clear, however, is the fact that movements of air here animate the body of the telegraph pole in such a way that seemingly dead nature experiences an uncanny resurrection. In Thoreau's description, air, wind, wood, industrial materials, and human sensory perception are part of a dynamic ecology of reciprocal attention, in which one element resonates with another, and any element requires the vibrations of the others to be perceived as such at all. Thoreau's wires and poles are endowed with what we typically associate with the human: the ability to listen and through attentive listening attune oneself to other forces; the capacity to perceive and to communicate perception. In this, Thoreau's wood and wire function as echo chambers, transducers, and amplifiers of the inaudible. As they empower the human subject to attend to what would otherwise exceed the

[5] Qtd. in Douglas Kahn, *Earth Sound Earth Signal: Energies and Earth Magnitude in the Arts* (Berkeley: University of California Press, 2013), 48.

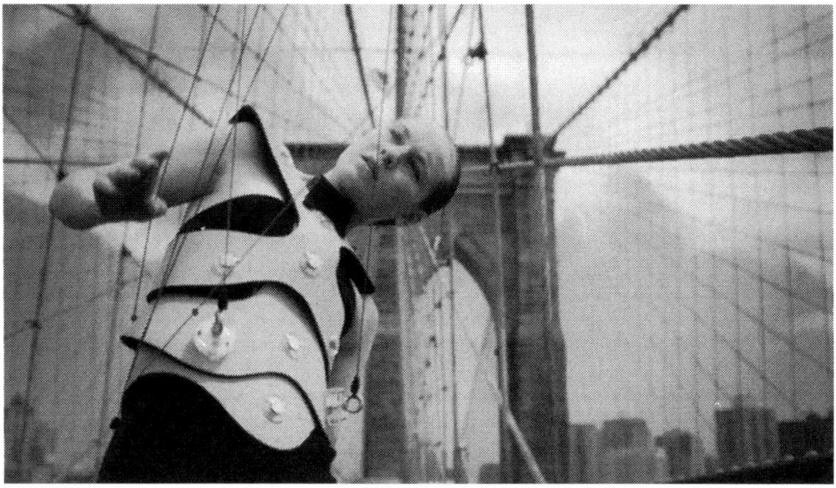

Figure 5.3 Di Mainstone: *Human Harp*, 2014.

wanderer's ear, technical interventions—the wireless's aeolian harp—transform the nature of hearing as much as they historicize supposedly timeless concepts of nature. Instead of obliterating natural phenomena, the resonant infrastructures of modern telecommunication in Thoreau's Deep Cut inaugurate new concepts of the natural. Nature 2.0. Air 2.0.

Following Thoreau's footsteps, numerous artists of the twentieth and twenty-first centuries have constructed imaginative wind harps in order to recode the relations of nature and technology, art and perception, the contingencies of climate and the weight of human civilization.[6] Perhaps the most striking effort to learn from Thoreau's experience of the Deep Cut can be found in a 2014 work by British sound and performance artist Di Mainstone, entitled *Human Harp* and first staged on New York's Brooklyn Bridge (Figure 5.3). Putting to work a relay of various feedback loops, *Human Harp* was initially based on sound recordings of the invisible and inaudible trembling of the steel cables of Brooklyn Bridge as caused by air movements, shifting weather patterns, and the streams of vehicular

[6] Think of Gordon Mohogan's 1986 *Long Aeolian Piano*, routing piano strings fifty meter in lengths through a piano's soundboard to transform air movements in Edmonton, Canada, into musical sounds. Think of Doug Hollis aeolian harps installed in 1976 along San Francisco harbor's pier 15 and 17 and inspiring numerous variations ever since. Think of Max Eastley's artistic research project *Aeolian Sound: An Investigation of Aeolian Phenomena*, in whose context different aeolian sculptures in urban and non-urban sites in the United Kingdom inspire new types of collaborations between musicians, visual artists, and scientists.

traffic. Digitally amplified, these sounds were then fed into a multitude of small speaker devices whose playing speed and volume could be controlled with the help of a mechanical coil system. In a third step, these speakers were attached to various cables of the bridge and its coils connected with individual wires and a unique corset to the body of a musician. By turning or bending her body, or by pulling on individual wires with her hands, this musician was then able to produce and probe a unique layering of frequencies and rhythms, a soundscape vaguely reminiscent of how Theremins in the first half of the twentieth century had once emancipated the production of music from the regiments of tonality.

Human Harp is remarkable for at least two reasons. First, in collapsing—like *The Visitors*—the boundary between the making of and the listening to sound, Mainstone positions the human subject as an interface between various technological and organic processes. The bridge's cables here emerge as extensions of the human body as much as the human body is being absorbed into the work's technological assembly. While the special effects of storage and reproduction media produce sensations of physical presence and immediacy, they at the same time allow the musician to uncover hidden aspects of reality—the acoustic unconscious, as it were—and restage the birth of music from the vibrations of and resonances between animated and seemingly inanimate matter. Second, and as importantly, *Human Harp*—at once recalling and recalibrating Kircher's experiments with aeolian music—sets up an ecology of listening in whose context marvelous sounds emerge as effects not simply of clever mechanisms that present art as nature's echo, but of techniques that fold material vibrations, air waves, technological media, and somatic processes into one dynamic. Sound here requires media and mediation to become sound in the first place, such that it does not appear far-fetched to argue that human-made structures such as the bridge tune into the movements of organic matter just as much as kinesthetic action attends to the vibrations of non-organic materials and suspends traditional concept of agential intentionality. *Human Harp*'s sounds result as much from bodies playing bridges as from bridges playing bodies. They generate unheard and wondrous tones for which the listener has no concept in spite and precisely because of the fact that she participates in their very creation.

Like the swells and sonic inflections Thoreau witnessed in the 1850s, Mainstone's perplexing sounds define hearing as a product of the heard: they require acts of listening to come into being. The performance activates new audible natures, reveals the vibrating aliveness of nonhuman objects, and thereby reshapes what it might meant to listen to begin with. Similar to more traditional aeolian harps, what's special about the special effects generated by acoustical media in Mainstone's work is not that they elicit, mimic, or affirm some preexisting notion of the real. What's special about them is their ability to make us curious about a world not solely governed by human-centered perspectives. What these sounds offer in the form of a promise is a world in which we can suspend narcissistic fantasies of self-management: a world we can openly encounter as one of unpredictable play and uncontrollable resonances, as one in which objects may have agency and physical matter can be experienced as much alive and in motion as human minds, bodies, and emotions.

Unseen Sounds

In the face of predominantly functionalist relationships to what philosophers once considered the natural world, Gernot Böhme argued in the 1990s that we experience nature today primarily in "atmospheric phenomena such as the wind, the weather, and the seasons."[7] Böhme's idea of the atmospheric conceptualized conditions in which materiality and subjectivity, the nonhuman and the human, non-intentionality and attentiveness, perceiver and perceived are interdependent. The cables that plug Mainstone's musician into the vibrations of the bridge, and the microphones, wires, and earphones that extend the functions of air and interconnect the musicians across the different spaces of *The Visitors*, serve as twenty-first-century stand-ins of the atmospheric. They enable resonant interactions that stress the mutual dependency of sound and listening. And as they redistribute the locations and actors of processes of perception, they recalibrate what we might call nature today. In both works, human-made materials take on, expand, and redefine the role of the elemental medium and atmospheric qualities of air. With this, both draw our attention to the fact that in our contemporary world neither wind, nor the weather, nor the seasons can really escape the footprints human populations have left on this planet. Yet in both works, the technological prostheses of air at the same time remind humans of their own limitations: their inexorable dependency on forces, vibrations, and movements that exceed visibility and control.

Pierre Schaeffer's experimental works and electronic compositions of the postwar period remain key to any modern discussion of sounds whose sources

[7] Gernot Böhme, *Atmosphäre* (Frankfurt am Main: Suhrkamp, 1995), 84.

elude visibility—the acousmatic. Unable to see each other, Kjartansson's musicians at Rokeby Farm index this tradition, even though they—as we will see in a moment—bend it into a different direction. The word itself is derived from the Greek "akousmatikoi" (ακούσματικοί), whose origins are typically associated with Pythagoras's practice to teach students while hidden behind a veil so as not to distract their learning with his visible presence and to focus their attention entirely on the spoken word and its rational force. Schaeffer himself advocated acousmatic listening practices as the most appropriate mode of experiencing avant-garde music. He championed modes of listening that reduced the perception of sound to the field of hearing alone and therefore had to eclipse whatever other sensory registers interfered with what Schaeffer called the sound object. Acousmatic listening understood how to focus on sounds solely for their sonic qualities. It reduced sounds to self-contained tones and eclipsed any possible awareness of or reference to whatever sources or locations that produced them.

Schaeffer's concept of the acousmatic has experienced a rich history in the realm of film theory and criticism, where it does useful work to elaborate on deliberate ruptures between the visible and the audible within a film's diegetic space. Acousmatic sounds in cinema typically expand and complicate narrative space. They create layers of signification, buttress or unsettle the filmic image, add (or at times subtract from) what viewers sense and know about the world invented by a certain film. But the legacy of Schaeffer's idea of the acousmatic, and his Husserlian plea for practices of reduced listening, certainly also continue to impact the work of contemporary composers and how certain forms of sound art are often idealized as quasi-automatic correctives to the visual's presumed excesses of abstraction, distance, control, and reason and its lack of interiority, immediacy, and immersion.

Consider, for instance, the work of Spanish composer and sound artist Francisco López. His 2007 recording *Wind (Patagonia)* provided a compilation of sounds captured in the desolate landscapes of southern Argentina, with its sonic textures meant to encourage acts of blind listening that explore environmental sounds as music, as putatively pure sound objects. *Wind (Patagonia)* continued sonic explorations that a decade earlier had led López to Costa Rica, where he compiled sounds from rain, waterfalls, insects, birds, and even plants into a complex acoustic environment meant to map the cycle of a day during the rainy season. When the resulting recording *La Selva* was performed publicly, López did not hesitate to ask his audiences to wear blindfolds in order to intensify the experience of "profound listening," that is, listen to nature's sounds without speculating about their possible sources. "It is my belief," López summarized his poetics of acousmatic purity,

> that music is an aesthetic (in its widest sense) perception/understanding/conception of sound. It's our decision—subjective, intentional, non-universal, not necessarily permanent—that converts nature sounds into music. We don't need to transform or complement the sounds. Nor do we need to pursue a universal and permanent assignment. It will arise when our listening moves away and is freed from being pragmatically and representationally oriented. And attaining this musical state requires a profound listening, an immersion in the *inside* of sound matter.[8]

The acousmatic in *La Selva*, like in *Wind (Patagonia)*, is meant to break the bonds of causality, the power physical matter holds over the listener's imagination and perception. In López's vision, to listen with your eyes wide shut is to restore the profundity of aesthetic experience: acousmatic listening here is designed to offer a training ground to cultivate attitudes that will protect the material world from the destructive effects of human willfulness.

Brian Kane's study of the acousmatic in twentieth-century theory and musical practice provides compelling reasons to approach the legacy of the Pythagorean veil in experimental music and art with some reservations.[9] In trying to restore some originary experience of and attention to sound in all its presumed purity, Schaeffer's method and, by implication, López's acoustic ecology tend to obscure the materialities, bodies, media, and technologies that help produce, propagate, and perceive the audible in the first place. The presumed purity and profundity of acousmatic listening in Schaeffer's and López's sense is a fiction and relies on profound ontological fallacies. It wants to isolate the mediation and listening to sound from any visual, tactile, and other sensory dimension because it considers the listener's sensing of sound as distractions to the aesthetic experience of music. Acousmatic listening in Schaeffer's and López's understanding eats its own tail. It pretends that sound and its media could live without body. It envisions aesthetic resonance without resonant vibrations and reverberations, without the inflections of timbre, space, expression, and propagating media such as air. And it precisely thereby neglects whatever allows sound to embody more than just a

8 Francisco López, "Profound Listening and Environmental Sound Matter," in *Audio Culture: Readings in Modern Music*, eds. Christoph Cox and Daniel Warner (New York: Bloomsbury, 2004), 87.

9 Brian Kane, *Sound Unseen: Acousmatic Sound in Theory and Practice* (Oxford: Oxford University Press, 2014).

sine wave, that is to say, whatever enables sound to become aesthetically vibrant and, to be frank, interesting in the first place.

The Visitors' approach to the acousmatic fundamentally differs from Schaeffer's practice of reduced listening and López's acoustic ecology. It does not consider vibratory movements, whether transmitted through the air or through the wires and membranes of other acoustic media, as mere distractions to the listener's aesthetic perception of music. And it does not think of the space in-between the origin of a sound and its listener as empty and uncontaminated, as a route of passage we best ignore when listening to music. On the contrary. In lockstep with Mainstone's *Human Harp*, *The Visitors* stresses the fact that sound and listening are messy and impure, whether or not air, space, and acoustical media render sources invisible, whether or not we seek to obscure the mediating properties of air. To place the band's eight musicians in various rooms of Rokeby Farm is not to blindfold them in López's or Schaeffer's sense. It is not to ask them to practice reduced listening and ignore the body of sound; nor to isolate the ear as the principle organ of attending to music. What it does instead is to expel any stubborn illusion about originary listening and, in stressing the process of technological mediation, to highlight the fact that music is never merely the ear's business anyway.

The concept of the acousmatic—listening to sound without its source in sight—draws awareness to what separates the space of sound production from the one of hearing, approaching this space as a site of aesthetic productivity, of intensified attentiveness. Like the aeolian harps of the Romantic age, the acousmatic appeals to aesthetic visions of sounds without authors and of music as something that may simply happen to us. Music might be all air and elude even in recorded versions our grasp, our desire to transform the passing of tones in time into stable objects. Yet, as we learn from *The Visitors*, nothing in the process of acousmatic listening, in listening to sounds unseen, needs to deny our effort to think of what separates source and ear as a neutral zone, an empty container, a transparent channel void of thickness, texture, vibratory inflections and refractions, even if this channel may only consist of the invisible molecules that make up the air.

Nowhere does this become clearer during the performance of *Feminine Ways* than when, halfway through the piece, Kjartan Sveinsson suddenly moves away from his grand piano. What unfolds is this: Sveinsson leaves his earphones behind, steps out of the frame of his screen, then immediately enters the screen that shows Davíð Þór Jónsson playing his grand piano, positions himself next to and eventually sings with him, revealing to the viewer that both musicians inhabited one and the same room all along and hence had no problem at all seeing each other. The isolation of sound from sight, we learn, is never as clean, clear, and complete as some of its theoreticians presume. Like Mainstone's dancer-composer-musician on Brooklyn Bridge, the performers of Kjartansson's *The Visitors* leave little doubt that music always needs media, including the medium of air, to come into being. Unlike the purists of acousmatic reduction, they make us think about the fact that matter other than the tympanic membrane is endowed with the ability to receive and attend to vibratory forces; that no music will resonate with listeners if it brackets the messy materiality of resonance; and that any effort to engage unseen sounds for the sake of intensifying the process of listening cannot afford to ignore the fact that sound never has a single source or origin in the first place. Sound's impurity, its lack of singular origins and authors, is music's wonder.

Architectures of Air

Music, writes David Toop,

> along with silence, is uncomfortably close to death—a shocking and shaping of the air, then suddenly gone. Sound may not excel in certainty until shaped with precision in certain forms of music, but both possess, in different ways, sublime attributes as metaphors or representations of all that is not solid: emotions, atmosphere, sensuality, passions of all kinds, the passage of time, abstract mathematics, memory, loss and mourning, nostalgia, the metaphysical, supernatural, spectral, unseen, unknown and uncanny.[10]

Music and sound resist capture, ownership, stability, closure. Under their sign, everything solid seems to melt into thin air. Like athletic performances, sound draws ephemeral figures in space.[11] Unlike sports, however, such figures in most cases even seem to elude visibility. They provide atmospheres that envelope listeners and blanket their perception. Sound's aerial life, its lack of solidity, makes it difficult at first to think of it as something we can inhabit, a resonant structure we can dwell in as if it were a home. Sound and resonance,

[10] David Toop, *Sinister Resonance: The Mediumship of the Listener* (New York: Continuum, 2011), 25.

[11] Hans Ulrich Gumbrecht, *In Praise of Athletic Beauty* (Cambridge, MA: Belknap Press, 2006).

as Toop stresses, are touched by the unhomely and uncanny. They, and their media, touch us as something secretly associated with death. As something sinister.

All air, music might be in cahoots with death. Yet few metaphors have experienced greater traction in the history of listening than the one describing sound and music with the help of architectural imageries. Most famous in this is no doubt Goethe's suggestion, expressed in his conversations with Eckermann in the late 1820s, to think of music as liquid architecture and architecture as frozen music. Following Goethe's lead, musician David Byrne quotes scientific studies suggesting that indigenous rock paintings in the Southwest often adorned cavernous sites with unique acoustical properties, echoes, and reverberations. For Byrne, such pre-Columbian intersections of art, sound, and the sacred indicate not only the extent to which places of communal gathering served as musical instruments themselves, but also the extent to which we should think of the relationship of architecture and sound as reciprocal: "Just as acoustics in a space determine the evolution of music, acoustic properties—particular those that effect the human voice—can guide the structure and form of buildings."[12] Sound might easily associate itself with the uncanny and sinister, but to think that audible or inaudible vibrations cannot but fail to provide spaces of dwelling and inhabitation misses the point. For sound in all its ephemerality does not simply echo in, fill, and modify given spaces and architectural structures. It also does not merely generate ersatz-dwellings in the air, however short-lived and intangible, frozen or liquid they may be. Instead, sound has the power to build worlds, generate spaces, and define environments on its own terms, and it thereby urges us to rethink the dominant nexus of space and living altogether. It asks us to expand dominant ideas of architecture and reframe what we tend to understand as dwelling.

Consider the work of British sound artist Dawn Scarfe, relocating Hermann von Helmholtz's nineteenth-century investigations of resonance and his development of the so-called Helmholtz resonator to the soundscapes of the twenty-first century (Figure 5.4). In preparing his groundbreaking treatise on acoustics, *On the Sensations of Tone as a Physiological Basis for the Theory of Music* (1863), Helmholtz devised a spherical brass device able to identify the frequency of certain tones in the surrounding world. With the device's smaller cone-like opening inserted into the scientist's ear, and the opposing larger

[12] David Byrne, *How Music Works* (New York: Three Rivers Press, 2012), 33.

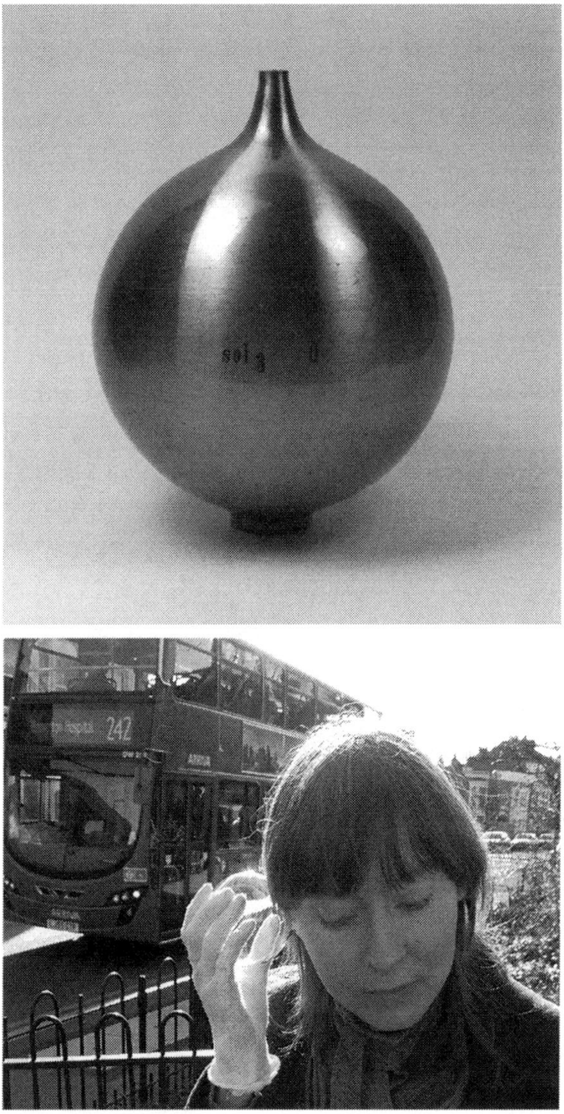

Figure 5.4 *Top*: Brass spherical Helmholtz Resonator from around 1890–1900. *Bottom:* Dawn Scarfe, *Through the Listening Glasses,* 2010.

opening directed away from the listener, the resonator dampened most sounds around the listener while it loudly featured frequencies resonating with the sphere's natural frequency: "Hence any one, even if he has no ear for music or is quite unpractised in detecting musical sounds, is put in a condition to pick the required simple tone, even if comparatively faint, from out of a great number of others. The proper tone of the resonator may even be sometimes heard cropping up in the whistling of the wind, the rattling of carriage wheels, the splashing of water."[13]

Scarfe's *Through the Listening Glasses* (2010) and *Listening Glasses* (2013), the first a fourteen-minute audio track, the second an interactive installation piece featured at the Botanical Gardens in Brussels, translated Helmholtz's scientific method into a project searching for the unheard music of today's human-made environments. In *Through the Listening Glasses*, Scarfe connected different Helmholtz resonators made of glass to binaural recording devices and took them for a walk: on London's Lea Bridge Road and South Mill Fields, through Vienna's city center. The resulting audio work, expected to be heard through earphones in order to experience the dynamic spatialization of recorded sound most effectively, offers a wondrous and uncanny map of different locations; its at times muffled or eerily hollow, at times shrill or curiously crisp, sounds shape a sonic world we at once recognize and not. During certain passages this world strikes the listener as extraterrestrial and subterranean, as an ambient landscape of washed-out, unidentifiable noise, so much so that it is difficult not to feel being inside and outside of something at the same time, to be shuttled back and forth between sensations of abstract interiority and embodied exteriority. Similar to listening and following the steps of Janet Cardiff's influential audiowalks, Scarfe's *Through the Listening Glasses* invites listeners to tap into the acoustic unconscious of certain places and vastly complicate their relation to the visual field. *Listening Glasses*, on the other hand, allows gallery visitors to avail themselves of various resonators and discover resonant frequencies—the unheard music—present in the exhibition space. To walk here is to experience one's own body and its technical extensions as instruments retuning the sounds of the real. It enables listeners to sound out and dwell in what evades the receptivity of the human eye: the invisible vibrations of the air.

According to Martin Heidegger, to dwell is to be situated in a certain relationship with existence characterized by acts of nurturing. Its basic character is to spare and preserve the things we build or cultivate; it is a staying with things:

> To be a human being means to be on the earth as a mortal. It means to dwell. The old word bauen, which says that man is insofar as he dwells, this word bauen however also means at the same time to cherish and protect, to

[13] Helmholtz, *On the Sensations of Tone,* Kindle file.

> preserve and care for, specifically to till the soil, to cultivate the vine. Such building only takes care–it tends the growth that ripens into its fruit of its own accord.[14]

Scarfe's dwelling in resonant sound takes us somewhere else. It is neither about a staying with things nor does it seek to cultivate and spare the materiality we employ to surround our mortality. To dwell in resonance, according to Scarfe, is to take the weak objectivity of sound, sound's ephemeral and aerial life, by its very words. Rather than to till the soil and grow vines, it invites the world to come to and enter our ears in all its unsteadiness, its uncanny alliance with disappearance and death. Rather than to reside in the solid and homely, it recognizes the inevitable passing of time and our perceptions as the very ground of human existence. Instead of preserving structures of old, it finds meanings, orientations, and fleeting moments of beauty in what exceeds our grasp. To dwell, when approached through Scarfe's listening glasses, is to be able to attend to and be struck by unexpected worlds and their movements, their conscious and unconscious vibrations. It is to understand one's body as no less a medium as the air that surrounds this body and that enables sound waves to travel to and through us. It is to encounter what we may cherish, care for, and love not at first but always as if it was at last sight.

Let me suggest to think of *The Visitors* and of how it understands to move the viewer in similar terms. Spread across different rooms of the mansion, the band members' aspiration is, not to preserve and cherish physical edifices, but to create and probe resonant sonic architectures that in all their fleetingness unsettle dominant understandings of dwelling. Rokeby Farm, for them, is as much a physical structure as a unique platform to approach infrastructures of solitary and common life through the lens and lesson, the conduits and vibratory forces of air. Whether they listen to unseen sounds or rework aeolian visions of tones without authors, the performers of *Feminine Ways* embrace what is ephemeral and sinister about sound as a condition of what it may mean to be human today. They inhabit resonant architectures of sound that urge us to rethink how human civilization has sought to prove its exceptionalism by considering nonhuman matter as dumb, passive, and entirely malleable only to produce inhospitable environments that now violently undermine humanity's sense of control, autonomy, and agency. We cannot succeed in caring for matters

[14] Martin Heidegger, "Building Dwelling Thinking," in *Poetry, Language, Thought*, trans. Albert Hofstadter (New York: Harper Colophon Books, 1971), 145.

of the world without being able to listen and be struck by the unexpected and wondrous. Rather than to dwell in continued fantasies of human sovereignty over matter, *The Visitors*—similar to Mainstone's and Scarfe's interventions—sculpts enchanting worlds of sound and air, ecologies of enchantment, in which subjects can rethink what it means to be a subject in order to make a case for their future.

6

Human, All Too Posthuman

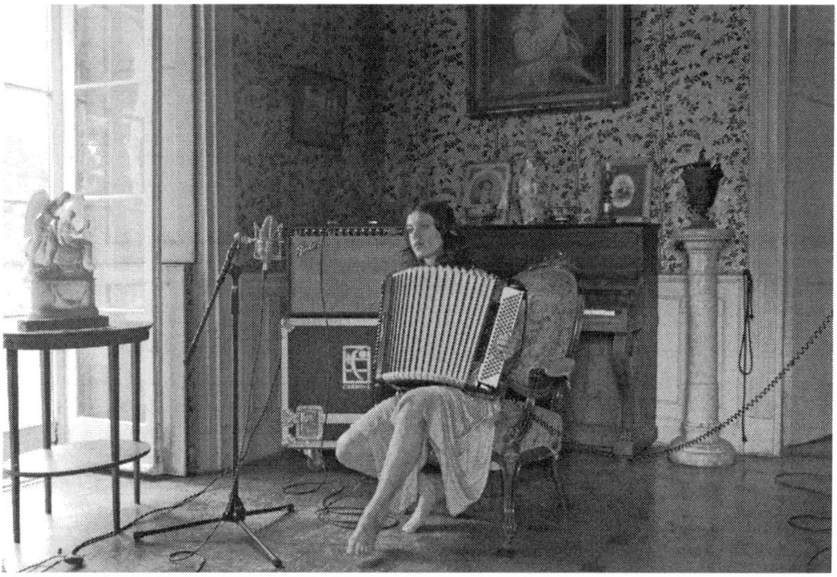

Figure 6.1 Ragnar Kjartansson: *The Visitors*, 2012 (Kristin Anna Valtýsdóttir). See also Plate 5.

The Limits of Articulation

As we have seen in the preceding three chapters, sound generates time and duration, as much as it takes up, inflects, unfolds in, and is space. Though a long history of Western thought wants to tell us otherwise, what we understand as the human voice clearly finds itself entangled in this logic of spatio-temporal becoming. Its sounds may aspire to master time and space and situate the subject as an autonomous agent. But no matter its media of articulation, propagation, and recording, no vocal activity can ever be abstracted from the contingencies of place and time, reverberation and resonance, context and history.

> As any squirrel, whale, or architect knows, sound is spatially affective, interacting uniquely with the materials, shapes, and configurations of the space in and through which it moves. The relationship between the human voice and the structure of the Greek amphitheater, or Gregorian chant and the Romanesque cathedral, or chamber music and the private salon is one of the mutual affective conditioning of one material by another, not simply the containment of musical matter by solid form.[1]

Enlightenment philosophers once considered man's ability to speak as a sign of (mostly) his humanity, of what distinguished him from the creaturely and allowed him to speak out in private and public realms on his behalf or on behalf of others. Unlike the animal or the mute world of mere things, the human was endowed with the aptitude of articulation: the capacity to interrupt the primordial flow of sound, sever sonic elements from each other, form individual syllables, announce separate words, sentences, and thoughts, and precisely thereby embrace the power of speech as a power to take hold over one's voice and unfurl its rationality. Condemned by jealous Gods to repeat the words of others over and over again, Ovid's nymph Echo clearly failed the bars later philosophers set up to prove one's humanness. To babble was to flunk the test of articulation, inviting sonic flux to wash over one's aptitude to mark distinctions. To solely resound the sounds of others was to forsake what it took to count as a subject and rational being, namely to own and execute the power of voice. Resonant voices, according to certain strands of Western metaphysical thought, don't qualify as voices at all.

1 Cox, *Sonic Flux*, 43–4.

Recent writing on sound and sound art has placed tremendous pressure on the coupling of voice, reason, and the human.[2] Whether it challenges this coupling from the standpoint of gender criticism as a historical marginalization of the female body or from a posthuman perspective as a precarious effort of centering history around the figure of the human, this writing tends to praise the spatio-temporal flux of sound as a medium to unhinge the legacy of the Cartesian subject and the fetishization of the (masculine) voice as an organ of individual sovereignty, human exceptionalism, and stalwart self-management. Properly understood, hearing—it is said—dispels any effort to install the subject as master over the dumbness of matter. For hearing, as Salomé Voegelin puts it,

> does not offer a meta-position; there is no place where I am not simultaneous with the heard. However far its source, the sound sits in my ear. I cannot hear it if I am not immersed in its auditory object, which is not its source but sound as sound itself. Consequently, a philosophy of sound art must have at its core the principle of sharing time and space with the object or event under consideration.[3]

Such a project, we are encouraged to conclude, does not only necessitate a focus on involved participation rather than detached mastery. It will also retune the rhetoric of autonomy we have come to invest into the human voice, the grounding of Western subjectivity, of individual agency and empowerment, in the reason of articulating sound. Sound's emphasis on immanence and the visceral asks us, in Holger Schulze's words, to recognize what is inherently WEIRD about dominant conceptions of the human (i.e., Western, educated, industrialized, rich, and democratic) and instead rethink our existence on this planet as that of humanoid aliens.[4]

The voice of *The Visitors*' Kristin Anna Valtýsdóttir no doubt lives up to such expectations. It is weird but not WEIRD. It has something profoundly alien, stretching dominant registers of the human voice to or even beyond their limits. Extremely high-pitched, it at times sounds metallic, as if funneled through some invisible vocorder, at other times like the screaming of a tortured

[2] For a selection of important texts, see Steven Connor, *Dumbstruck: A Cultural History of Ventriloquism* (Oxford: Oxford University Press, 2000) and "The Strains of the Voice," in *Phonorama: Eine Kulturgeschichte der Stimme als Medium*, ed. Holger Wölfle (Berlin: Matthes and Seitz, 2004), 158–72; Adrianna Cavarero, *For More than One Voice: Toward a Philosophy of Vocal Expression*, trans. Paul A. Kottmann (Stanford: Stanford University Press, 2005); Mladen Dolar, *A Voice and Nothing More* (Cambridge, MA: MIT Press, 2006); Joshua Gunn, "Gimme Some Tongue (On Recovering Speech)," *Quarterly Journal of Speech*, 93 (2007): 361–4; and Noonan, *Echo's Voice*.

[3] Voegelin, *Listening to Noise and Silence*, xii.

[4] Holger Schulze, *The Sonic Persona: An Anthropology of Sound* (New York: Bloomsbury, 2018).

creature, a creature torturing mainstream listeners. It's not that Valtýsdóttir, when contributing to the singing of *Feminine Ways*, simply stresses what Roland Barthes called the voice's grain, that is, uses her vocal organ to communicate the materiality of the speaking or singing body, to exceed the operations of signification and the meaning of words, and to displace the "fringe of contact between music and language."[5] For Valtýsdóttir's strained voice, in refusing or simply ignoring the exigencies of "normal" articulation, makes us wonder whether her sounds have a body in the first place; whether the signifying force of words is never more than a bleak illusion; whether what we hear from her corner of the room at Rokeby Farm is music or language at all.

A colleague, when viewing *The Visitors* at the Frist Art Museum in Nashville in 2016, called Valtýsdóttir in private communications the Minnie Mouse voice. His point was not to belittle its aesthetic value, but on the contrary stress the extent to which Valtýsdóttir uses her voice to push against human-centered visions of speech and song—to emphasize the extent to which her voice makes us think about sound as mediated and special effect. Valtýsdóttir's voice emancipates sound from the normative templates of articulation. In all its artificiality and extraterritoriality, its tinny and overtly shrill monotone, it seems to defy anything associated with resonance or reverberation, yet in so doing it makes us rethink the exclusion of Echo's sounds from the lexicon of the human as well. At once comical and deeply disturbing, Valtýsdóttir's voice slides across the distinctions Western humanism invented to install man as superior to mere matter and the creaturely. Like the voices of Echo and Minnie Mouse, she sounds out a world in which the alien and artificial, man-made machines and cyborgs, the synthetic and the non-organic often turn out to be more human than the human itself.

Resonance is typically associated with acts of listening to or being affected by the vibrations of others, be they human or not. To resonate with something, at both the physical and the more metaphorical level, is to be attuned to the movements of something different, strange, unfamiliar, or other. It is to allow this other to take up shop in the house we call our own, or to attend to—to stretch oneself toward, to attach oneself to—what at first may strike us as incompatible. A resonant voice, by extension, is one able to echo, assimilate to, chime in, swing along, absorb, and be absorbed by other sounds, whether they emanate from the voices of others or simply present audible and inaudible

[5] Roland Barthes, "The Grain of the Voice," in *Image—Music—Text*, trans. Stephen Heath (New York: Hill and Wang, 1977), 181.

vibrations of matter. Resonant subjectivities, in playing with and being played by other sounds, become and perform to be something other. They entangle, attach themselves, and fall into what exceeds their ability, willingness, or zeal to mark and make distinctions, to articulate flux and flow into discrete units.

While recent writing on sound often praises such a decentering of the subject as a welcome step into a posthuman future correcting the hubris of anthropocenic positions, other critics may find considerable fault with the mimetic qualities and unconditional hospitality of resonant subjectivity. To emulate Minnie Mouse, to forfeit the logic of articulation as a base of establishing both personhood and rational communication—all this, for this different set of critics, means no less than to throw out the human with its bathwater. How, they ask, can we challenge contemporary politics of precaritization, populist nativism, and systemic racism, how can we argue for greater equality in gender relations, how can we empower marginalized communities and populations, if we relinquish the concept of voice and concomitant ideas of individual agency? How can we re-envision present and future if we desire to disarticulate the human voice and in so doing possibly undo the ground of politics altogether?

This chapter lends our ear to various projects that tackle the ground in-between these seemingly irreconcilable positions. Rather than siding with one stance or the other, and rather than endorsing an unbending perspective of either/or, their intervention is to shift the entire territory of current debates about voice and subjectivity, the coupling of articulation, agency, and power. In straining the voice to complicate (rather than erase) our understanding of subjectivity and articulated communication, and in meshing the force of listening into dominant concepts of voice, these interventions provide rich material to think of Valtýsdóttir's Minnie Mouse sound, not as the end of the human, but a promising beginning of a new and different one amid the inhospitable times of our present.

Inhale | Exhale

The voice must leave the body to gain shape and volume and be identified as such. We must lose our voice, exhale it, to gain or regain it. Paraphrasing and extending the work of Steven Connor, Brandon LaBelle writes accordingly:

> The voice is inside and outside in one and the same instant; it is spoken and heard, in the head of the speaker, as vibratory sensation and expelled breath,

> and as signifying gesture, as communicable message. Thus, we recognize our voice only as it leaves us, only at the moment of its articulation, as that initial paradox identified by Connor—when it rides on the wind to return to us, as if from another. The voice is in control and out of control; it reveals agency in the words spoken, which form commands, pleas, and invitations, and, in turn, it dissolves agency, leaving the speaker depleted, helpless, and unable to conjure words so as to enter conversation.[6]

Whether they originate from Tibet, Mongolia, Sardinia, or elsewhere, traditional throat singers are known to unsettle this paradox of vocalization. They collapse the distinction between what is in and out of control, between moving the voice into the world and absorbing the world back into the body, between speaking and hearing. Belying common expectations that resonance is all about consonance, harmony, and pleasantness, the eerie sounds of traditional throat singers result from a singer's trained ability to manipulate resonant vibrations as they travel from the singer's lung past his or her vocal cords. Throat singing—often also identified as overtone singing—makes use of various resonant cavities along the channel of human exhalation (the pharynx and larynx, the mouth) in order to produce either individual sounds rich in overtones or multiple pitches at once that are different from the frequency produced by the singer's vocal folds at a given moment. Ears unaccustomed to the haunting tones of accomplished throat singers will find little to recognize their orchestration of resonance as the product of human voices, let alone as music or song. A throat singer's sounds appear at once too far inside and too much detached from the body to appear as a form of vocalization. They cut across what might be perceived as being under and as being beyond control to live up to common expectations about vocal agency, and in this way they fold into one what is typically relegated to interior and exterior spaces and separated as vibratory force and signification. What is perhaps most uncanny and paradoxical about the resonant sounds of throat singers is the fact that they brush against what causes critics to speculate about the paradoxical features of the human voice.

We must lose our voice to gain it, exhale subjectivity to inhale the world. First Nations artist Tanya Tagaq (Figure 6.2) not only counts as one of North America's most gifted vocalists, but has understood how to blend traditional Inuit throat singing with the itineraries of contemporary experimental and

[6] Brandon LaBelle, *Background Noise: Perspectives on Sound Art* (New York: Continuum, 2006), 105–6.

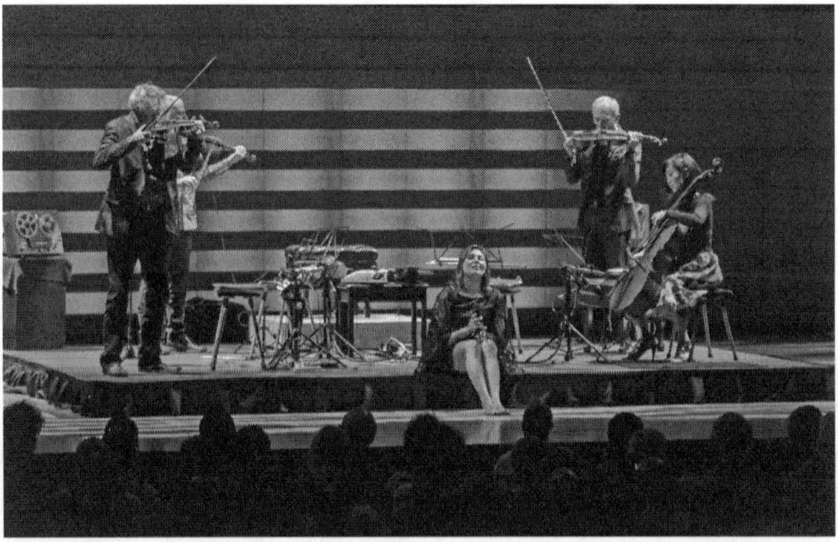

Figure 6.2 Tanya Tagaq, performing together with the Kronos Quartet as part of their Fifty for the Future project, at the Royal Conservatory, Toronto. May 25, 2016.

avant-garde music. Born in Cambridge Bay (Iqaluktuuttiaq), Canada, Tagaq is known not just for exploring throat singing as a solo practice rather than following the conventional form of the female duet, but even more so for lending her resonant voice to international collaborations with artists such as Björk, Matthew Barney, Jessy Zubot, and, perhaps most remarkably, the Kronos Quartet. After first performing with the members of the Kronos Quartet at Vancouver's Chan Centre for the Performing Arts in 2006, Tagaq was invited to compose a piece for the Kronos's Fifty for the Future project in 2015–16. The resulting piece, entitled *Sivunittinni* (The Future Ones), asked the Quartet's members to develop unique techniques to transpose Tagaq's multi-tonal signing into sounds teased out of string instruments—no small feat at all. "Kronos has gifted me the opportunity," Tagaq described the process, "to take the sounds that live in my body and translate them into the body of instruments."[7] Though using traditional staff lines, the score developed for this piece uses various visual marks that defy conventional models of notation, reflecting the fact that Tagaq actually does not read music. Just as importantly, verbal cues repeatedly prompt musicians to perform tactile gestures that, in spite of their descriptiveness, leave considerable room for interpretation: circles with slashes call for "white noise,

[7] http://kronosquartet.org/fifty-for-the-future/composers-detail/tanya-tagaq.

no pitch, with very light bow, like an inhaled breath," solid or open triangles on different staff lines indicate "light bow pressure, to sul tasto, like an exhaled breath."[8] In *Sivunittinni*, string instruments thus do not simply seek to mimic the resonant sounds of Inuit throat song; they do whatever possible to resonate with and become the breath that enables Tagaq's articulation of multiple pitches. They at once inhale and exhale and as a result, like Tagaq herself, mimetically unsettle the place of what we want to consider voice in the first place.

We have come to understand the concept of mimesis primarily as an act of representing the real, of capturing and copying the world in the form of an image. In the process, the word's alternate meaning, meant to describe processes of vibrant reciprocity between subject and object, processes of being moved and carried away, of being deeply affected and in fact infected by something that exceeds our understanding, of blurring boundaries and becoming other, has gone somewhat underground.[9] *Sivunittinni* unearths and puts into play this other meaning of mimesis. The composition calls for strings to be infected by Tagaq's voice, violins to become vocal chords and breaths, performers to echo and be carried away by sounds that themselves echo and resonate with the body's cavernous spaces.

Snapping the Voice

When one is listening, writes Jean-Luc Nancy, "one is on the lookout for a subject, something (itself) that identifies *itself* by resonating from self to self, in itself and for itself, hence outside of itself, at once the same and the other than itself, one in the echo of the other, and this echo is like the very sound of its sense."[10] Tagaq's resonant and multi-pitched vocal practice and the way in which the Kronos Quartet's strings seek to swing with and chime into Tagaq's strange sounds, recall what Nancy describes as listening's search for a subject, the echoing of something other in the self. The experimental work of opera singer Julianna Snapper (Figure 6.3) over the last decade is driven by a similar aim, yet approaches it from the opposite side of the spectrum, thus further illustrating what it may mean to stage one's voice as an echo chamber of something strange,

[8] https://kronosquartet.org/images/uploads/Sivunittinni_-_Performance_Note.pdf.

[9] Michael Taussig, *Mimesis and Alterity: A Particular History of the Senses* (New York: Routledge, 1993).

[10] Jean-Luc Nancy, *Listening*, trans. Charlotte Mandell (New York: Fordham University Press, 2007), 9.

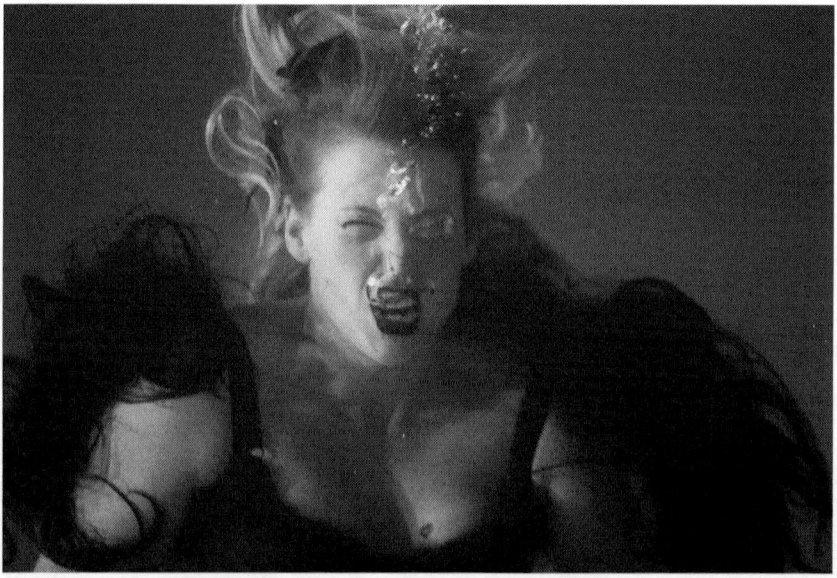

Figure 6.3 Juliana Snapper.

something that exceeds normative understandings of the human, something whose resonant weird-ness—like Valtýsdóttir's cartoonish sounds—casts light on the WEIRDness of dominant understandings of the human.

Snapper's experimental vocal work intends to push the voice to and beyond its limits. To strain it amid its medium such that it may no longer be able to resonate, reverberate, sound, and articulate anything at all. To snap it. In *Judas Craddle* (2005–7), Snapper explored the extent to which she could undo the operations of the vocal cords by hanging upside down, with a light arch to the back, from the bars of a jungle gym. To sing under such strained conditions was not simply to reveal how mainstream opera conceals the labor of the singing body in all its physicality. It was to place this body in such a position that it could no longer serve the purpose of what is typically understood as the process of articulation—a position in which the singer's vocal chords were taken over by and began to echo some unknown to which no identity can be attributed:

> Through rigorous experimentation Snapper located the point at which she, as a singer, lost control, allowing her voice to take over as an autonomous, driven, and determined entity. Her own voice hastened her to places where her knowledge of singing and her artistic imagination could not take her. In other words, she discovered that allowing the physicality of her instrument, rather

> than prewritten instructions or preconceived ideas, to dictate the sound of her performance led her to new possibilities.[11]

Unlike Tagaq's, Snapper's voice in *Judas Craddle* barely makes any recognizable sounds at all and hence is unable to cause other objects to vibrate along. What she profiles is a voice that swallows, cancels, or disassembles its very ability to resonate. A voice that gets stuck in the body precisely because it has been reduced to being all body and instrument. And yet, in doing so, Snapper—like Tagaq's *Sivunittinni*—opens a space in which her listeners cannot but reconsider normative associations of voice and body, articulation and the human. Snapper's performative annulling of vocal resonance and expressivity—like Tagaq's collaboration with the musicians of the Kronos Quartet—unravels facile couplings of agency and voice as much as it complicates our thinking of sounds as audible vibrations easily traveling from self to other, interior to exterior, sender to receiver.

This dislodging of the human comes to the fore most dramatically in Snapper's repeated efforts to perform individual solos underwater and compose and stage entire operas for aquatic settings. Unlike the work discussed earlier in this book, Snapper's underwater performances are not about testing the properties of water to convey sound at higher speeds, altered pitches, and in ways that explore the possibilities of non-cochlear listening. For Snapper's audiences typically remain seated as listeners outside the pools, tubs, and glass tanks she uses to immerse her body in water to activate her vocal organ. What they are asked is not only to hear Snapper's voice as it is being refracted, contained, and rendered strange while it travels from water through air to their (dry) position of listening, but to witness her struggling and occasionally succumbing to what water does to human actors seeking to use their voice underwater. In her *Five Fathoms Deep Opera Project* (2008), Snapper used pool settings, water tanks, and bathtubs in different performance locations to explore the lure of sirens and shipwrecks in opera history, deliberately allowing different aquatic environments to inflict painful obstructions to her effort to communicate her singing of Baroque and Wagnerian arias. It remains an open question whether Snapper's singing underwater is less masochistic in nature compared to Odysseus's cunning act of tying himself to the mast of his boat to enjoy the beauty of the Sirens' songs, or his rowers' use of wax to escape the lure of their seductive song. What simply

[11] Eidsheim, *Sensing Sound*, Kindle file.

matters for our context here, however, is the fact that Snapper, in restraining her voice, snapped normative ideas of articulation, communication, and beauty and that in hampering the voice's ability to resound Snapper derailed common associations of voice, resonance, and controlled forms of agency.

Sound, Snapper teaches us, does not exist in a vacuum: the material qualities of media that help propagate sound matter distinctly. More importantly, however, what her work accomplishes is to challenge the vibrational resonance of vocal articulation and in this way open a room in which we may readjust the normative coupling of voice and human sovereignty. In Kafka's famous retelling of the ancient myth, silence turned out to be the sirens' most deadly weapon in face of Odysseus's instrumental rationality, his willful use of his own body as a tool of accumulating pleasure. As if seeking to translate Kafka into the medium of sound art, Snapper's work silences possible sounds so as to reveal what is illusive about any vision to control one's body like a tool or object. She strains her body and drowns her voice's resonance to feature the mutual entanglements of the human and the creaturely—the limits of strategic reason and the curious pleasures that reside beyond dominant expectations of audible melos, beauty, consonance, and harmony. To sing means to be forever on the lookout for a subject. It is to resound with the echoes of otherness at the at once pleasurable and painful costs of dodging the burdens possessive individualism imposes on self and subject.

Hysteria

The (male) medical profession of the eighteenth and nineteenth centuries was obsessed with pathologizing as hysterical voices that seemingly lacked the power of control and reasonable articulation. Drawing on the Greek term for the uterus and the Greeks' belief that its potential movements would eventually suffocate a woman's breath, physicians, psychologists, and psychoanalysts thrived in considering the emotional excess associated with hysteria a "woman's disease," a disorder of femininity that in its telling by men threatened the foundations of bourgeois society.[12] As much as the many alleged symptoms of female

[12] In 1883, French physician Auguste Fabre did not hesitate to state that "all women are hysterical and . . . every woman carries with her the seeds of hysteria. Hysteria, before being an illness, is a temperament, and what constitutes the temperament of a woman is rudimentary hysteria." Qtd. in Elaine Showalter, "Hysteria, Feminism, and Gender," *Hysteria Beyond Freud*, eds. Sander L. Gilman, Helen King, Roy Porter, G. S. Rousseau, and Elaine Showalter (Berkeley: University of California Press, 1993), 287.

hysteria served male ideologues as evidence to push back the claims of modern suffragettes, so they in the eyes of Sigmund Freud and Josef Breuer provided the engine to develop psychoanalysis's central methodology, the "talking cure." Understood as a profound disturbance of female sexuality, hysteria needed an entire new science and medical practice to guide afflicted women to regain breath, speech, and narrative and hence live up to what men themselves had defined as the normative expectations of being a whole human. Hysteric bodies resonated with forces that hampered the resources of human agency, autonomy, and meaning-making. Talking cures desired to talk sense into what escaped meaning. They unstrangled the voice, delivered it from the pathological grasp of resonance, only to remake it according to the image Western thought had developed about the proper shapes and forms of the human over many a century.

(Female) hysteria has ceased to be part of the medical language. It no longer figures as a derogatory scientific concept of lack designed to marginalize female voices by means of reducing women's bodies to resonant echo chambers of unchecked forces. In late-twentieth-century feminism, however, hysteria took on a new role in order to challenge continued asymmetries of gender and—transposed into an affirmative concept—cut across the legacy of hierarchical dualisms of reason and resonance, articulation and echo. What once described a mental disorder whose diverse symptoms prohibited the development of reliable forms of subjectivity was now reclaimed not only to reveal the systematic oppression of women in Western societies but also to rattle the very framework of binary ideas, norms, and values that buttressed this oppression and othering in the first place. Recall Hélène Cixous's famous 1975 intervention encouraging Medusa to laugh rather than being struck by silence about her men-destroying powers:

> Listen to a woman speak at a public gathering (if she hasn't painfully lost her wind). She doesn't "speak," she throws her trembling body forward; she lets go of herself, she flies; all of her passes into her voice, and it's with her body that she vitally supports the "logic" of her speech. Her flesh speaks true. She lays herself bare. In fact, she physically materializes what she's thinking; she signifies it with her body. In a certain way she inscribes what she's saying, because she doesn't deny her drives the intractable and impassioned part they have in speaking. Her speech, even when "theoretical" or political, is never simple or linear or "objectified," generalized: she draws her story into history.[13]

[13] Hélène Cixous, "The Laugh of Medusa," trans. Keith Cohen and Paula Cohen, *Signs* 1.4 (Summer 1976): 881.

Similar to how the Black Lives Matter movement has transformed Eric Garner's "I can't breathe" into a powerful chant of political activism, so did the former suffocating of female voices during presumed states of hysteria become an outcry against regimes of power that considered female bodies as mere hosts of processes that exceeded her grasp. To reclaim hysteria in the late twentieth and early twenty-first centuries meant to rally against ideological infrastructures that effectively relied on ostracizing the hosting of affects and echoes as void of what it might take to assume the position of a human subject and agent. It drew bodies and vibrant materialities into history rather than pushing them outside.

Such a quasi-Nietzschean act of transvaluation, of revaluing naturalized values, norms, and perspectives, is at the core of Camille Norment's 2015 sound art installation and performance *Rapture*, which represented Norway at the 56th Art Biennale in Venice (Figure 6.4). Installed in the Nordic Pavilion, the glass of its windows partly splintered, its broken frames piled up on the interior concrete floors, *Rapture* involved musical sounds played by Norment on a glass harmonica, an instrument originally invented in the eighteenth century to produce tones whose utter purity could trigger otherworldly—rapturous—effects in the bodies of their listeners. Yet rather than elicit experiences of musical transcendence, Norment's ambition was to probe ambivalent constructions

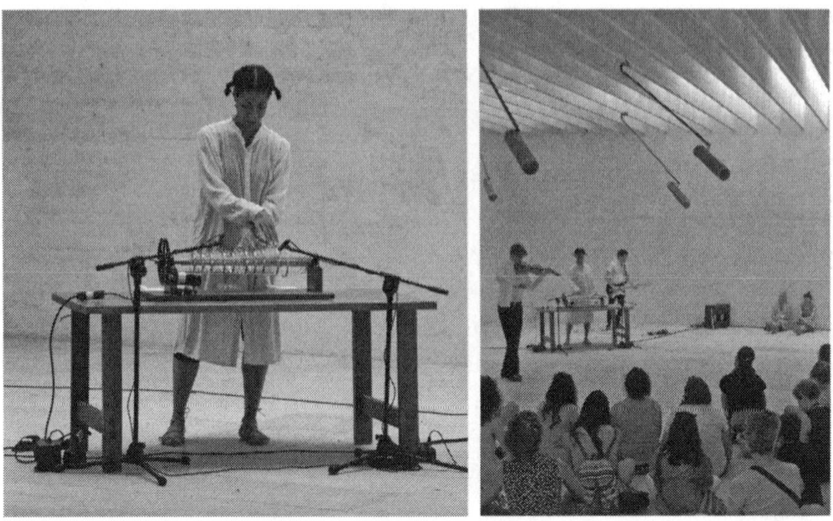

Figure 6.4 Camille Norment: *Rapture,* 2015. Performance at the 56th Art Biennale in Venice.

of audible vibrations as both a source of physical sickness and psychological disintegration, of presumed female hysteria, and as media of bodily and mental healing and rejuvenation. "I am interested," Norment stated,

> in how music has long been used to facilitate both the forging and transgressing of cultural norms. Sound permeates all borders. Throughout history, fear has been associated with the paradoxical effects music has on the body and mind, and its power as a reward-giving de-centralizer of control. Recognized as capable of inducing states akin to sex and drugs, music is still seen by many in the world as an experience to be controlled—especially in relation to the female body—and yet it is also increasingly used as a tool for control under the justification of war.[14]

Rapture investigated the paradoxes of sound—the fears and hopes associated with music's ability to decenter concepts of sovereign and self-managing subjectivity, of voice, agency, and control—by exploring the infectious power of resonant interactions: the way in which the touch of fingers may activate potent vibrations and sounds from rotating glass; high-frequency sounds may impact non-organic matter and possibly break them in the process; musical instruments may attune to each other through feedback, reverberation, and echo; and sounds may have the power to set human bodies into motion and usher the subject beyond the brinks of self-control and strategic action.

"Glass," writes Norment, "is a formation of sand, water, and time. The sound of glass is at once beauty and noise, and aligns itself with both the analog and the digital. Glass is physically present, but its transparency renders it nearly invisible. Invisibility is an existential reality for society's 'unseen.'"[15] Performances of Norment's *Rapture* in Venice aspired to render this dimension of the unseen audible, visible, and sensible. When glass harmonicas were invented, the purity of their sounds was associated with the music of angels as much as with sinister forces harming—like Homer's sirens—the integrity of the male body. Known for his theory of animal magnetism, German physician Franz Mesmer (1734–1815) employed the instrument in his effort to heal female patients from various maladies and afflictions—melancholia, hypersensitivity, deliria, uncontrollable weeping, partially obstructed vision—that Freud a century later lumped into the category of female hysteria. In her Venice performance, Norment—clad like a doctor in white robes—invited musicians such as David Toop, Håvard

[14] http://www.e-flux.com/announcements/camille-norments-rapture/.
[15] http://www.norment.net/highlights/solo-performance-series/.

Skaset, and Vegar Vårdal to improvise on various instruments to the sounds of her glass harmonica, causing Vårdal in particular to whirl through the gallery space like a dervish as he played his fiddle, mesmerized by sounds whose striking absence of overtones and depth approached the seemingly disembodied sounds of digital sine waves. Her fingers on the rotating glass thus set off unpredictable echoes and reverberations, uncannily interconnecting the vibrations of organic and non-organic matter, of architectures and bodies, of flesh and mind, of the audible and the inaudible.

Similar to Di Mainstone playing and being played by the Brooklyn Bridge, performances of Norment's *Rapture* offered an ecology of attention in which sound hovered along and constantly challenged the border of the perceptible. The apparent purity and precision of her sounds rang strange and extraterrestrial, as if emanating from sources and regions that own no name; they seemed to come from deeply ambiguous places, neither entirely material nor completely immaterial, located somewhere in the folds of the visible and perceptible field. Yet rather than dwarf the listening subject in the mode of the sublime, *Rapture* enabled a joyful decentering of human subjectivity, a suspension of human-centered forms of perception, a reciprocal attuning of vibrant matters, that could not but energize the listener's curiosity. Though *Rapture* made no effort to hide highly technological processes and various techniques of sound production, in the end it too—like many other projects featured in these pages—set up an environment for the wondrous, for sounds that in placing listeners beside or outside of themselves caused awe for that which exceeds given concepts, expectations, and judgments.

Rather than to reiterate stories told by men about female hysteria, then, *Rapture* offered sounds and vibrations meant to collapse conventional hierarchies of mundanity and excess, restraint and flow, program and improvisation, the angelic and the demonic. To be hysterical here becomes the new normal, the very ground from which any distinctions and hierarchies unfold. To resonate with forces, energies, and intensities that fray the harness of the modern subject here is to recognize that we can never own or secure the workings of what we call subjectivity—that we can never be more than on the lookout for a subject and its voice. At once beauty and noise, the sounds of glass in Norment's *Rapture* feature resonance, not as the nemesis of (male) reason, agency, and identity, but on the contrary as a switchboard of vibrational energy that asks us to complicate these concepts and rethink the stubborn historical coupling of voice and agency. As the harmonica's putatively angelic sounds send Vårdal demonically whirling through the audience, Norment set up an environment in which the movements of material receptivity, of attuning and attaching the body to uncategorizable vibrations, belong at least as much to what it means to be human as to our ability to distinguish ourselves from the creaturely and nonhuman through acts of articulation and self-assertion.

The Resonant Human

It is easy to think of Kristin Anna Valtýsdóttir's shrillness in *The Visitors* either as a celebration of the creaturely, the cyborgian, and the posthuman or as a deliberate strategy of unleashing hysteria against its own history, of performing the weird and decentered to expose what Schulze calls the WEIRDness of normative expectations about gender identity and subjectivity in the West. In this sense, for Valtýsdóttir to sing like Minnie Mouse is to entangle herself with a cinematic cipher, a special effect, an animated machine whose anthropomorphic features testify to whatever is and has gone wrong about the human project. Like Tagaq and Snapper, she strains her voice so as to draw attention to this voice's material base, its grain, while at the same relinquishing the ability to assume full control over her sounds. She cancels, in the very act of articulation, what modern thought in the West has associated with the process of articulating words, namely the assertive self-differentiation of the human from the nonhuman. And like Norment, Valtýsdóttir's voice brings into play resonant infections to upend histories that considered the hysterical as a negative in response to which Western thought has repeatedly sought to posit instrumental reason and human self-management—the tenets of anthropocenic agency—as a positive. She plays at being beyond control so that both her collaborators and her audiences recognize the rapturous as the vibrant ground of all being and becoming.

It is easy to think of Valtýsdóttir's voice as a foray into what transcends or precedes the human—and hence as an event that, in our disastrous twenty-first century, either exposes to view or in fact undoes the calamitous impact visions of unconstrained humanity have imprinted on this planet's elemental media. Yet maybe there is no need to throw out the human baby entirely with the contaminated bathwaters of our age, and instead simply understand the interventions of Valtýsdóttir—and, for that matter, of Tagaq, Snapper, and Norment—as important efforts to reassess triumphalist visons of human agency, as efforts to probe alternate formations of human subjectivity attuned to the fact that nonhuman creatures, things, bodies, and materials have some agential powers as well and that the nonhuman may own a voice, a mode of articulation that the human can no longer afford to ignore. Nowhere does Valtýsdóttir's individual performance articulate this perhaps more urgently then when we witness her snugged into her armchair cupping her hands over her earphones so as to attend intently to the sounds of her co-performers and feed her own voice back into the resonant network of Rokeby Farm. Her gesture iconically couples acts of listening and singing, ear and mouth. It defines attentive hearing as a precondition for sonic expressions, folding the operations of the voice into one's ability to listen both to the sounds of others but also and as importantly—as we can learn from Tagaq's or Snapper's experiments with resonance—to the incommensurable and surprising vibrations that reside within our own bodies.

The act of listening has not fared well in cultural criticism and political theory. It has been associated with passivity and mere receptivity, has been sidelined as infested by emotion and affect, has been relegated to the realms of privacy and interiority and seen as a mere addendum in philosophies trying to define the virtues of public life, the dynamics of political leadership, and the condition of the human. We celebrate eloquent speakers, admire crafted verbal arguments, fear the rhetoric of demagogues, and abhor the lies of powerful politicians. We think of speech in both public and private realms as the decisive medium to negotiate differences, carry out conflict, act in the world and through such action constitute our human identity, define the contours of the social, and navigate the realms of the political. Whereas speech has thus figured central in dominant understandings of human agency, power, and autonomy, listening has experienced a fate similar to that of matter. If not entirely ignored, it has been engendered as inactive and submissive, void of agential powers and incapable of bearing any kind of ethical, political, or public force. Unlike speech, writing, and reading, listening tends to be taken for granted, as if it could have developed outside history and culture and without technological mediation, as if good listening wouldn't take as much training, discipline, and effort as good speaking, reading, or writing. Listening is aligned with nature in contrast to speech's culture, in spite of the fact that listening, as Kate Lacey has passionately argued, "is at the heart of what it means to be in the world,"[16] to engage in social relations, and in fact act as a political being.

[16] Kate Lacey, *Listening Publics: The Politics and Experience of Listening in the Media Age* (Cambridge, UK: Polity Press, 2013), Kindle file.

Whether they strain the voice or make the body's sounds resonate in ever-new ways, the acoustical experiments we encountered in this chapter all push against historical efforts, in the name of vocal agency, to sideline listening as passive. They not merely seek to insist on the rather banal insight that there is no sound, no speech, no song without its listener. In all their experiments with non-normative sounds and forms of articulation, they all instead hope to complicate what we understand as voice—its mediums, it locations, its modes of being in the world. Throat song, underwater opera, and Minnie Mouse singing do not take place in a vacuum. They feature the human voice as an echo chamber of other forces and subject it to resonant interactions that exceed the itineraries of controlled articulation. In this, they identify the voice as an organ whose operations essentially require prior acts of listening, of attending and attuning to other sounds, of absorbing and being absorbed by various vibrations, in order to project sound into the world in the first place. There is no voice, no form of articulation, they suggest, that does not understand how to host what is strange and foreign, including the strangeness and foreignness of the very body that speaks or sings. There is no voice without a body—and no body that resides beyond the contingency, flux, infectious rapture, and vibratory dynamics of all matter.

Our times, during a period of self-destructive human agency on Earth, are in dire need to rethink the claims matter in all its vibrancy may have on our present and future. We will not succeed in this if we fail to transform how we have come to think about the reach of the human, and how we have afforded agency and voice to certain agents on this planet at the costs of declaring others as voiceless, passive, inarticulate, and void of agency. Triumphalist visions of future human geo-engineering, meant to reverse the calamities of the Anthropocene with even more technology, human interventionism, and market forces, might as much miss the point as defeatist positions asking for radical withdrawal and celebrating human self-destruction as the best matter on Earth can hope for.

The matter of cinema, the chemical base of its medium as much the world of objects it captured, inspired critic Siegfried Kracauer in the mid-twentieth century, without letting go of his hope for a better future, to think of the non-identity of objects as a domain able to make us reassess the reach of human thought, identity, and agency: "Guided by film ... we approach if at all, ideas no longer on the highway leading through the void but on paths that wind through the thicket of things."[17] Kracauer, who experienced world historical catastrophes as an exile of Nazi Germany

[17] Siegfried Kracauer, *Theory of Film: The Redemption of Physical Reality* (Princeton: Princeton University Press, 1997), 309.

on his own body, could not help but consider the future relation between the human and the world of matter as a vabanque game, a tension-ridden entanglement of bodies, technologies, voices, claims, and agential powers with a fundamentally unpredictable outcome. The decisive wager in this game was for us to learn how to think as if we were extraterrestrials: to recognize the resonance between the alien in us and the incommensurable language and voice of the material world. Bringing Kracauer's thought into conversation with that of another exiled film theorist of the early twentieth century, Rudolf Arnheim, Jennifer Fay concludes: "Out of this thicket of raw material and cinematically induced sensations, a new . . . love for inert, physical reality may emerge—one that even Arnheim admitted had the potential to bring about new, radical thought and therefore strategies of survival in a damaging world erected on an earth that persists beyond us. Thinking like extraterrestrials, we may be able to form nonbinding attachments to hostile places we have never called home, to care for an environment that exists in our absence, and to cultivate 'disinterested' identification with an unaccommodating ecosystem undetermined by 'previous preference.'"[18]

In its call to approach the materiality of our world like extraterrestrials, Kracauer's thought remains relevant to our times. His proposition is not to give up humanistic standpoints. It is to cultivate a humanism—a resonant form of subjectivity and human agency—attuned to the voices, claims, and agential powers of the world of things. Kracauer anticipates a reformed humanism whose extraterritoriality, entanglement with the other, and refusal to endorse visions of human exceptionalism continues to drive the experiments of performance artists such as Tagaq, Snapper, Norment, and Valtýsdóttir. As weird as the latter's voice may strike the audience of *The Visitors*, it bears testimony to *The Visitors*' humanist aesthetics and politics of hospitable coexistence. Valtýsdóttir's sounds explore the pleasures, pains, and obligations of forming attachments to places we may call home amid times of self-inflicted catastrophe. Recalling Kracauer's vabanque game of history, Valtýsdóttir's voice indexes what it takes to pay tribute to material environments whose vibrations may very well outlast the very thing we have come to call the human subject. None of this will save the planet from the damage humans have done to it already. What it extends, however, is the aesthetic promise of a resonant humanism whose focus on the redemption of physical reality includes the care for voices and bodies of matter that exist, and will exist, even in the absence of human agents.

[18] Jennifer Fay, *Inhospitable World: Cinema in the Time of the Anthropocene* (Oxford: Oxford University Press, 2018), 198.

7

Silence

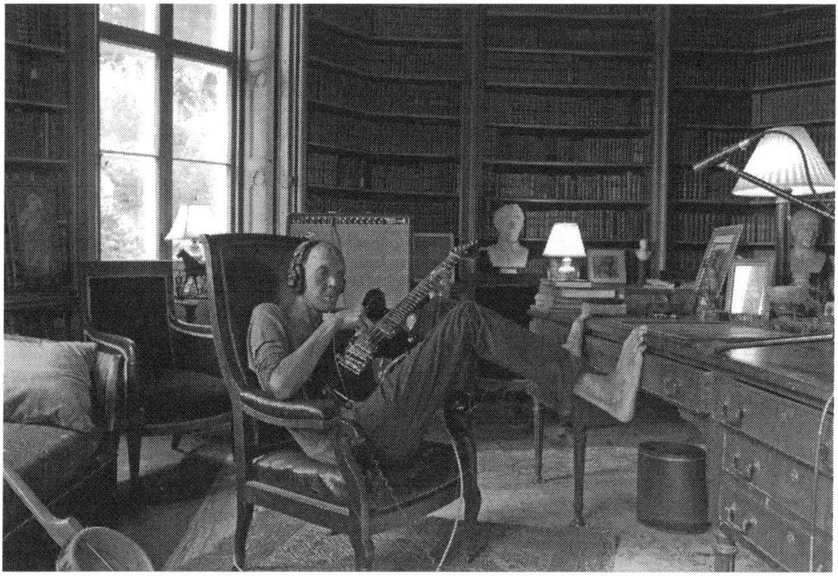

Figure 7.1 Ragnar Kjartansson: *The Visitors*, 2012 (Shahzad Ismaily). See also Plate 7.

A Diamond Heart

Ragnar Kjartansson's *The Visitors* performs and reframes a story of loss. It enters the void of a failed relationship with sound, song, and music, not in order to make the shipwreck of human attachments move splendidly forward again, but simply to cope with agony by probing how to move with its burden. In Kjartansson's aesthetics of "Scandinavian Pain,"[1] no one speaks of a victorious rekindling of shell-shocked affects. To survive in spite and with them, to reclaim some sense of futurity, of mobility, of vibrancy in their very face, is all we can hope for. This may at first sound like very little, but in the end might be more than most are ever privileged to accomplish.

The ventriloquizing of pain and agony is perhaps Kjartansson's most recognizable artistic signature. In his single-channel video *God* of 2007, Kjartansson repeats the phrase "Sorrow conquers happiness" hundreds of times over thirty minutes (Figure 3.2). Dressed like a 1950s entertainment singer, Kjartansson tackles his slogan each time with ever-so slight variations in intonation, volume, and posture, often looking straight into the camera as if to draw the audience right into his space, swaying his hips in what increasingly becomes a blurring of the lines between the serious and the humorous. As James Gibbons writes: "The gloom about sorrow's ascendency is joined, through repetition and song, by a countervailing, almost tender sense of consolation. In its inviting, oddball, post-Romantic way, 'God' spans the grandest of contradictions: the intertwined truths that sorrow holds dominion over our happiness, but we might yet find ourselves liberated from its thrall."[2] To repeat with slight differences, to repeat such differences incestuously, is to outperform the very gloom and melancholy that holds sway of a subject in pain. Pushing against sorrow's imperial grasp, it assumes, in Gibbons's words, the status of liberation theology. It embraces the durational—non-climactic time—as a house in which hearts hardened into diamonds may learn, ever-so cautiously, to beat and resonate with the world again.

Kjartansson's work over the last decades appears to be driven by a categorical imperative demanding sound and music at all times and costs. His performances,

1 So the giant pink neon motto at the entrance of Kjartansson's mid-career retrospective at the Hirshhorn Museum in Washington, DC, October 2016–January 2017, curated by Stéphane Aquin in collaboration with the Barbican Centre, London.

2 James Gibbons, "Liberation Theology: Ragnar Kjartansson's 'God,'" *Hyperallergic*, December 10, 2016, https://hyperallergic.com/343176/liberation-theology-ragnar-kjartanssons-god/.

including *The Visitors*, refuse dramatic high points and push narrative arcs as much as possible to the sidelines. They last longer than most viewers and listeners are typically prepared to spend during normal gallery visits. Their internal form and structure—like the modernist grid in Rosalind Krauss's reading[3]—present a song's material as entirely knowable, as something that neither contains dark secrets nor allows anyone to get lost in any way. And yet, Kjartansson's play with pattern, repetition, and variation transforms the time of aesthetic representation into a space reluctant to contain the movements of sound and vibration. It leaves little doubt that there is no logical boundary, no final gesture that could separate any artwork from what surrounds it, no necessary reason why things could not go on and keep vibrating forever. The fake Viking ship that in Kjartansson's 2013 performance *S.S. Hangover* (Figure 7.2) sailed back and forth between two adjacent docks of Venice's Arsenale, carrying a sextet that played highly iterative

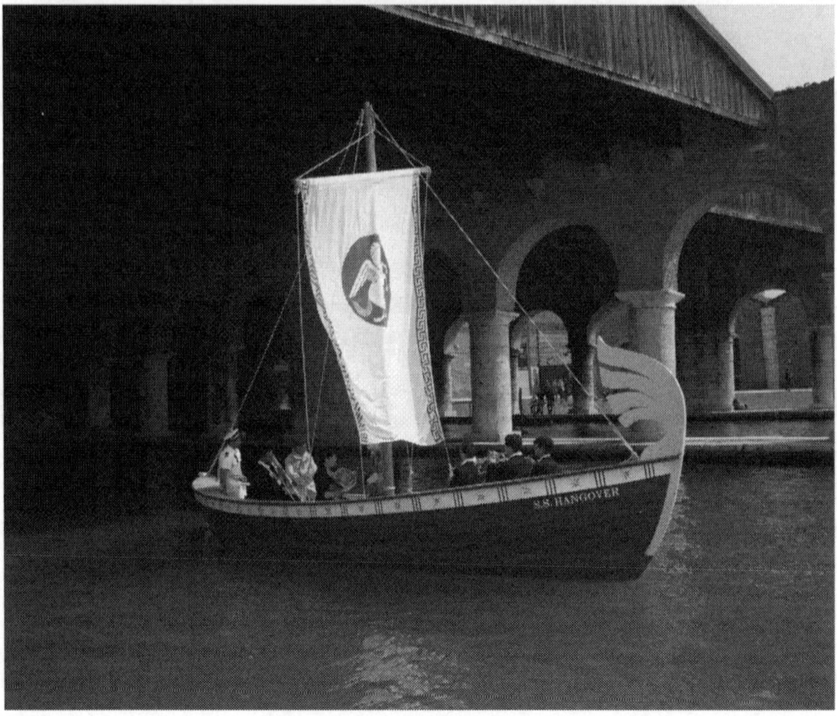

Figure 7.2 Ragnar Kjartansson: *S.S. Hangover*, 2013–14.

[3] Rosalind E. Krauss, "Grids," in *The Originality of the Avant-Garde and Other Modernist Myths* (Cambridge, MA: MIT Press, 1985), 8–22.

brass music composed by Kjartan Sveinsson, had no obvious reason to end its course after the prescribed six hours. It could have stopped anytime precisely because nothing in the music's formal design restricted it from not going on for hours on end. Kjartansson's work is at once minimalist and maximalist. Its sounds, in all their repetitiveness, push up against conventional frames of a "work" not only to bleed into the everyday, but due to their suggested endlessness—the sense of infinite vibrations—to provide what it takes to keep the world's sorrow, pain, despair, trauma, and melancholy at bay. Repetition liberates. Liberation needs repetition—revolutions in the word's original astronomical sense—in order to find its foothold in life.

In spite of Kjartansson's quasi-manic devotion to enduring sound, however, it takes little to notice that *The Visitors* is quite mindful about the possibility of silence around its edges—about silence as a condition from which sound's vibrations emerge and may collapse back into. The piece starts with all nine screens muted until a technician will turn each room's recording device on one by one. The first musical sound we hear is a very gentle strumming from Kjartansson's guitar, soon to be accompanied by softly sung lyrics, as if querying whether and how sounds can kick off time and make space vibrate before some of the other musicians chime in from their distant locations. Just as the 136-bar E♭ major drone in the prelude of Richard Wagner's *Das Rheingold* allegorized the creation of world and music from nothing, so does Kjartansson's initial guitar play and voice enact the emergence of sound from the domains of silence. No one is likely to find rings of gold in the Icelander's bathtub, but undulating waters serve both Kjartansson and Wagner as media to acknowledge the void against whose backdrop neither sound nor music may exist. Throughout the 64-minute recording of *The Visitors*, the musicians will repeatedly approximate or in fact fall back into moments of silence, only to then reboot their play, restart their musical weaving, sneak and gently goad sound out of the void of silence. And in the end, after eight screens have already turned dark again and as we see Kjartansson and friends walk down the estate's lawn and depart from any microphone able to record their sounds, the piece approaches the very silence from which it emerged more than an hour earlier—even though this final silence, like the quietness after the last notes of Gustav Mahler's *Ninth Symphony*,[4] is

[4] Vera Micznik, "The Farewell Story of Mahler's *Ninth Symphony*," *19th-Century Music*, 20.2 (Fall 1996): 144–66.

pregnant with meaning and vibrates with memories of earlier departures, transformations, and transfigurations. It is not mute at all.

The sounds of Kjartansson's *The Visitors* emerge from the silence that ensued after a painful breakup. The performance seeks to reclaim some form of vibrancy by hosting the words of Ásdís Sif Gunnarsdóttir's poem, by embracing its echoes as a touchstone of a renewed sense of movement, connectivity, and community; it thus explores what it takes to listen into and sound out the shapes of silence. Rather than simply wanting to fill up the void with music as if loss and absence had been mastered, *The Visitors* reconstructs resonant interactions from a void that appeared to obstruct the very possibility of resonance. Its point is to probe how to let in and live with what is missing and lost—to invite the void to resonate with future presents, however paradoxical that may sound.

Resonance is a concept many typically charge with profound presentism.[5] What is absent or missing cannot resonate, they argue. What is mute or lifeless cannot serve as a source of resonant vibrations. Voids, they add with good reasons, obliterate the possibility of echoes and reverberations: they can neither touch upon and resound in our lives, nor can they ricochet anything sent its way. Voids swallow the very possibility of material vibrations infecting other matter to vibrate, as much as they incapacitate any effort to develop affective or empathetic relationships. We bellow into a space of silence, but all we experience is that there is no there there. Any aesthetic or ethics of resonance, it therefore seems, cannot but fare badly to attend to what refuses contact and communication, what is absent, what has been lost or has disappeared. Though voids may cause the human subject to tremble with fear, despair, or humbleness, their nature is to consume the very work of resonance and thereby amplify our sense of individual or even planetary loneliness.

But then again, even without taking on the stance of a resolute Derridean or a hardcore Deleuzian, there are many good reasons to claim that no present is ever as present, and no absence, loss, or lack as void as we like or fear it to be. The complexities of consciousness and signification, the interplay of perception, memory, and anticipation, the entanglements of sensation, knowledge, delay, deference, and desire, the temporality of spatial forms and the spatial dynamics of temporal figurations—all suggest that neither time nor space occur in the singular, that no one can ever hope to dissolve one entirely into or successfully insulate one from the other. Resonant vibrations travel from past to future and

5 Rosa, *Resonance*, 47–83.

possibly trigger all kinds of other pulsations along their way in what we call the present. They live in and across time without ever culminating in one singular moment or in one isolated location. Affects are messy and promiscuous, at heart illogical, and therefore seep into spaces and times we may believe to be off the grid, unresponsive, out of grasp. The silence seemingly created by loss inevitably rings, in however distorted or internalized form, with the echoes of grief, pain, recollection, and trauma of those left behind. No present exists without the resonance of what is absent. No absence may be perceived as such without the sounds and images we engage to map what we understand to be our present.

Kjartansson's *The Visitors* knows a thing or two about the dialectics of sound and silence, presence and absence. It offers an intriguing model to host what has been lost; sound out voids that pierce and haunt the present; listen to what has become silent and transform spectral echoes into something that overcomes sensations of stasis, arrest, and vertiginous loneliness. What's more, the piece aspires to redeem the work of resonance from false assumptions that deny sound and vibrant matter to transcend their particular moments in time and enter the historical. The installation descends into a presumed void of resonant interactions—a diamond heart of darkness—in order to patiently recuperate what it takes to reanimate the seemingly lifeless. In thus recognizing the historicity, temporality, and durational multiplicity of resonant attachments, *The Visitors* encourages us to think through what resonance in art, aside from mapping the holes left by the end of romantic entanglements, can do today to address the absences produced by historical traumas, violence, and repression. It is to two projects that translate Kjartansson's personal liberation theology into a deeply political aesthetic that we turn our attention in the pages to follow. As these interventions explore the silent voids of traumatic pasts, they invite the lives of the dead and murdered to resonate with the movements of the living. In a third step, we then draw on another recent installation by Ragnar Kjartansson, *Death Is Elsewhere* (2019), in order to reformulate our question as to the relation of silence, loss, and resonance in light of humanity's ever-more foreseeable self-extinction.

The Rumbling of the Vanished

International news media in 2018 circulated the term "Havana Syndrome" to describe alleged acoustic attacks in 2016 and 2017 leading to serious psychosomatic health problems among the staff of the American and Canadian embassies in Cuba. Though the Cuban government swiftly denied any activity, numerous theories flourished for some time about the weapons possibly used to carry out such assaults, whether they resulted in what the afflicted described as the hearing of strange discordant noises or as unpredictable vibrations of the entire embassy building. Others have tracked the history of allegations and counter-allegations, and poured over numerous assessments by medical experts.[6] What solely matters for our context here is the eagerness with which reporters, pundits, and the larger public thrived on the idea of sonic warfare, exploiting the invisible nature of acoustical vibrations both to propagate wild suspicions without empirical proofs and fan fears about the power of what exceeds sight to manipulate our minds.

As a political event, the Havana incident testified to the extent to which Cold War imaginaries continue to dominate the twenty-first century. From the perspective of contemporary media theory and aesthetics, on the other hand, the developments in Cuba serve as a stark reminder that—no matter what really happened in Havana—sound and resonant vibrations indeed can be highly manipulative, can cause pain and injury, can be used to declare war on the senses, and exceed the kind of playfulness, autonomy, and freedom we typically associate with the realm of aesthetic experience. As we have seen before, resonant vibrations score high on the charts of cultural theorists eager to rethink reigning epistemological, ethical, and aesthetic concepts from the standpoint of sound so as to develop new ecologies of mindfulness, sustainability, and mutuality. However, whether merely fictional or not, the Havana Syndrome reminds us of the fact that sound's vibrant matter can certainly also fuel subliminal choreographies of power and be used to enlist the senses in effective schemes of despondency. Resonance blurs the boundaries between subject and object, but such a blurring can cause pain as much as it can cause pleasure; it may open our minds and emotions to the matter of the world as much as it may enslave bodies and obliterate the possibility of thought. Resonance, in short, does not automatically afford "good object" status. Whatever ethics or politics we seek to attach to it is a matter of ambivalence, of contention.

The recent work of Mexican conceptual, video, and performance artist Teresa Margolles investigates this political dialectic of resonance, the ambiguities of

[6] For a meticulous reconstruction of the political fallout, see Adam Entous and Jon Lee Anderson, "The Mystery of the Havana Syndrome: Unexplained Brain Injuries Afflicted Dozens of American Diplomats and Spies. What Happened?," *The New Yorker*, November 19, 2018, https://www.newyorker.com/magazine/2018/11/19/the-mystery-of-the-havana-syndrome.

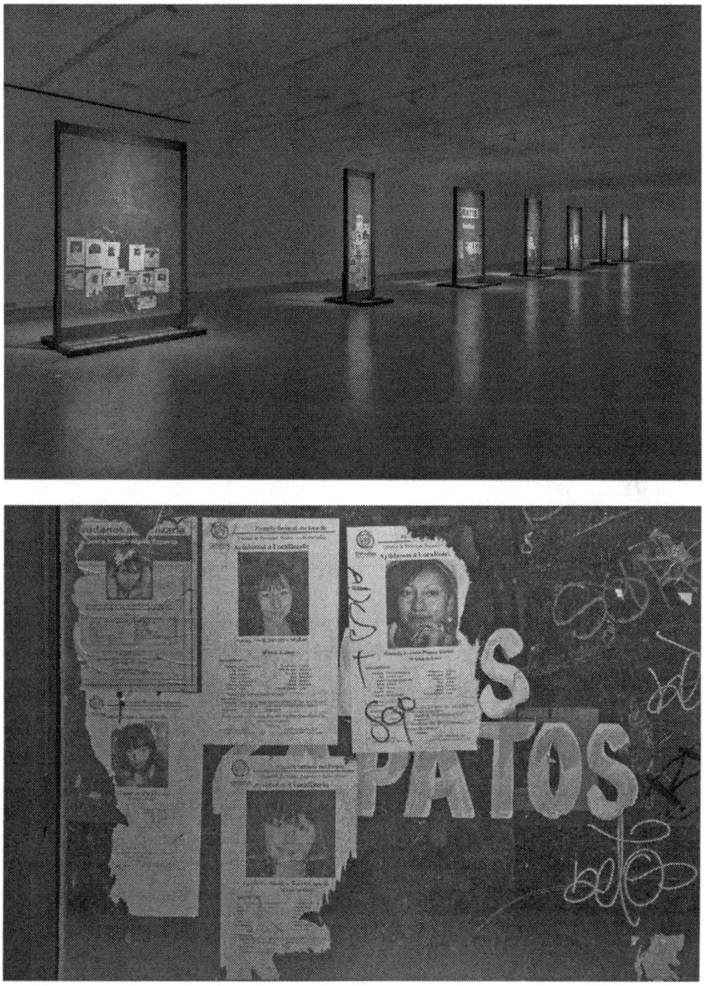

Figure 7.3 Teresa Margolles: *La Búsqueda* (The Search), 2014.

audible or inaudible vibrations, the tension between the heard and the silent or silenced. Her installation *La Búsqueda* (Figure 7.3), first shown—like *The Visitors*—at the Migros Museum of Contemporary Art in Zurich in 2014, recalls the tragic fate of countless women in Central America victimized by the violence of narcotrafficking in Mexico at least since the 1990s. The work revolves around a whole series of posters once displayed on shop windows in downtown Ciudad Juárez, one of the major hubs of the drug trade, a stone's throw away from El Paso on the other side of the Mexican-US border. These posters were once meant to aid the search for women whose sudden disappearance could not be explained:

victims of, in most cases, rampant femicides that targeted and continue to target female students and young mothers who migrate to the region to earn a living in foreign manufacturing plants set up along the border. Though many of these women had been missing for many years already, their search notes remained visible in public spaces for many years after their disappearance, as haunting signs communicating the presence of a painful absence. Instead of relying on the work of digital reproduction and virtualization, Margolles in preparation for her installation decided to acquire entire shop windows covered with such posters, reframed their glass, and then assembled them in a straight line in the gallery setting. What audiences thus encounter is a porous wall of glass, a gallery of memory stemming against the brutalities of history—a tribute to the vanished that upholds the claims of unlived lives against the amnesia and violent frenzy of narcotrafficking.

Things get more complicated, though, once audiences come to realize that resonant sounds are central to *La Búsqueda* as well. Subtly attached to each of the window panes, special loudspeakers play sound recordings from a train running through Ciudad Juárez. The rhythmic noise they emit is in low-frequency range, less meant to be heard with our ears than felt with our skin and guts. A steady rumble. A roll and reverb that has no real direction, yet imprints itself uncomfortably onto the perceiving body. Margolles's sounds cause her glass walls to vibrate ever-so lightly. The trembling is not sufficient to break the panes and thus rip the posters into pieces, but it certainly is sufficient enough to prompt the viewer's imagination to anticipate the possibility of such a shattering collapse of the entire structure. Similar to the afflicted staff workers in Havana, the quivering of the glass in Margolles's "aesthetics of commemoration"[7] can be sensed as a strange throbbing of the air and surrounding atmosphere, a contraction, pulsating, and shivering of space causing considerable perceptual unease. Unlike the Havana Syndrome, however, *La Búsqueda* invites the audience to identify possible sources of this throbbing, traverse its extensions actively with our bodies, and probe the hold it may have on our perception. In this Margolles situates the viewer as a searcher true to the installation's very title. Nothing is designed to really shatter here. There is little chance, however, that audiences will leave the gallery without having experienced a profound disturbance of their frames of attention, without carrying along into the future

[7] A term used by Anthony Downey to discuss earlier work by Margolles, in "*127 Cuerpos*: Teresa Margolles and the Aesthetics of Commemoration," in *Understanding Art Objects: Thinking Through the Eye*, ed. Tony Godfrey (London: Lund Humphries, 2009), 104–13.

sensory memories of something unknown, yet deeply transformatory, bound to happen—ghosts of a past that knock against the suspension of time.

It is useful to think of what audiences of *La Búsqueda* encounter as "unsound" in Steve Goodman's sense. Unsound, in Goodman's understanding, refers to inaudible sounds, be they below or above the threshold of what the typical human ear is able to perceive. Goodman's notion of unsound, however, also describes sounds we do not yet recognize as such, that exceed our knowledge and categorization—unheard-of sounds, sounds-to-come, and sonic effects on the body for which no concepts exist yet. Unsound, in this usage,

> describes the peripheries of human audition, of infrasound and ultrasound, both of which modulate the affective sensorium in ways we still do not fully comprehend. In its negative connotation, unsound aptly describes the colonization of inaudible frequencies by control. But most important, unsound also names that which is not yet audible within the normal bandwidth of hearing—new rhythms, resonances, textures, and syntheses.[8]

The trembling of *La Búsqueda*, somewhat audible, yet mostly felt through our skins, points toward such a breaking of the vessels of time, not in order to subjugate our senses to wicked regimes of power, but instead to recall, honor, and carry forward that which was silenced by violent histories and never allowed to live. However, given the fact that there are many different ways to interpret the train's noise rumbling through the installation space, such recall is ambiguous. It ushers us into speculative realms. Does this sound recall the prolonged travels that brought migrating women to Ciudad Juárez and work as so-called "maquiladoras" tax-sheltered companies exporting their products to other countries? Does it communicate their hope for economic improvement or the sorrow about leaving families behind? Or both at once, blended into a challenging hybrid? Is it meant to provide a sense of forward movement, the hope for better times to come? Or of painful standstill and circularity amid false promises of progress? Does it signify the tragic journeys of numerous women now vanished or the feverish mobility of narcotraffickers moving their products and, in playing up to long-standing macho images, victimizing women they see as easy targets?

There is no single answer to any of these questions. The trembling of Margolles's window panes, their unsound, generates tremendous unease among the

[8] Goodman, *Sonic Warfare*, 191.

perambulating viewer of the installation. Call the effect ghost-like. Call it the Ciudad Juárez syndrome. Resonant sound here rattles against the sovereignty of visual perception. It destabilizes the boundaries between cochlear and non-cochlear forms of audition, and it also erases clear lines between different temporalities, rhythms, and moment in times. While we anticipate the relocated glass walls to shatter any moment, and while we cannot but fail to synthesize the different meanings suggested by the rumbling of the train, *La Búsqueda* in all its unease offers its audiences ample space to imagine lives never lived. It does not do so to put to rest the tension between mobility and standstill, the visible and the absent, transitory time and death, noise and silence. On the contrary, the installation's unsound instead urges us to recognize that any sense of futurity rests on our ability to carry memories of the vanished forward and create some space for their absence.

Resonance—the glass's hazardous vibrations imprinting themselves onto the bodily tissues of the viewer—here approaches us as something precarious, as spectral. And yet, to those who follow Margolles's invitation and examine the posters on display up close with caution and care, the resonant unsound of the installation urges us to embed the void of the silenced in any future present lest we want to perpetuate and become complicit in the violence of the past. Unable to make sense of the air's ghostly and perilous vibrations, what remains for us is to explore how to resonate with the lives of the vanished and continue to listen for possible echoes in the unmarked graves of histories. To keep searching. On and on.

Res·o·nant Voids

When it opened to the public in 2001, Daniel Libeskind's Jewish Museum in Berlin stood out as an architectural structure whose intricate design, like no other before, was able to meet the difficult task of memorializing the victims of the Holocaust in postwall Germany (Figure 7.4). In the words of the architect's studio, the design rigorously followed three decisive insights and principles:

> It is impossible to understand the history of Berlin without understanding the enormous contributions made by its Jewish citizens; the meaning of the Holocaust must be integrated into the consciousness and memory of the city of Berlin; and, finally, for its future, the City of Berlin and the country of Germany must acknowledge the erasure of Jewish life in its history.[9]

[9] https://libeskind.com/work/jewish-museum-berlin/.

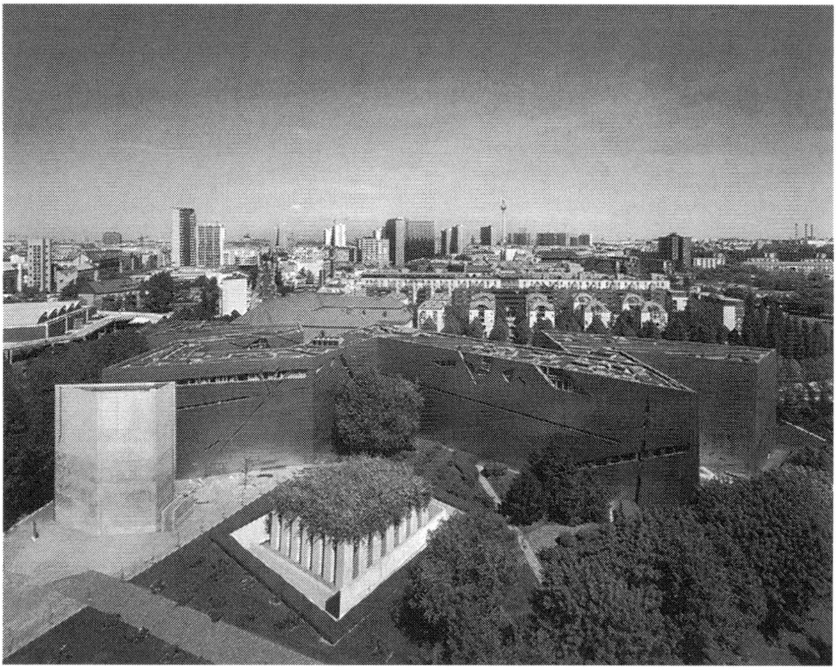

Figure 7.4 The Jewish Museum Berlin, designed by architect Daniel Libeskind.

The building's zigzagging ground plan; its canted angles and three interior axes of movement; its material expression of ideas of historical continuity and rupture, dead end and hope, extermination and exile, memory and survival—all these, in the eyes of the majority of the building's initial critics, offered a successful architectural language to represent what eludes representation—the enormity of the crimes of the Holocaust—while providing apposite spaces to memorize the contribution of Jews to German culture even before exhibition objects actually populated the museum's galleries.

One of the most discussed features of the Libeskind building are its numerous architectural voids: concrete shafts that slice through the structure, are neither heated nor air-conditioned or lighted, and interrupt at both a cognitive and kinesthetic level the flow of possible movement through the building. Most of these architectural volumes are not accessible to the museum's visitors. They are marked off by walls painted in black or are only discernible from what the design identifies as "void bridges." Their physical emptiness is meant to represent in visible, albeit paradoxical form, the void left by the annihilation and expulsion of Europe's Jews during the Holocaust—an emptiness no politics of remembrance, restitution, or reconciliation could ever fill again. In this, these voids, in Libeskind's understanding, serve as spaces defining the limits of what the building in its function as a museum can and should exhibit. Their task is to operate as a conceptual reminder that no act of memorization can ever reverse the course of traumatic histories.

Of the few accessible voids, one houses a permanent installation by Israeli artist Menashe Kadishman, entitled *Shalekhet—Fallen Leaves*. Cut from round iron plates, more than 10,000 faces cover the void's ground, most of their mouths wide open in what haunts visitors as silent screams, their facial features abstract and yet varied enough to confront us with the unresolvable tension between recalling the Holocaust's unfathomable extend of state-executed murder and heeding the fact that pain, suffering, and death is always individual. Two additional voids, including a connecting pathway, were made available in fall 2017 to host temporary exhibitions. Covering a combined floor space of more than 350 square meters and affording ceilings twenty-four meters in height, this double void presented as its first exhibit a multimedia installation, entitled *res·o·nant*, by German conceptual and light artist Mischa Kuball. *Res·o·nant* was on display between November 17, 2017, and September 1, 2019, with its location within the larger geography of the building so central that no visitor entering the museum through its underground pathways and slanted axes was able to miss it. It is to Kuball's effort to aesthetically probe the resonance of Libeskind's voids that we now turn.

Res·o·nant's visual and sculptural components consisted of two rotating projectors casting light fields on the shafts' walls and floors, their shapes emulating the architectural footprints of Libeskind's voids, their ongoing movements drawing the eye to wander across the building's brutal concrete in new and unexpected ways (Figures 7.5 and 7.6). Kuball's site-specific play with moving light recalls the work of the early twentieth-century avant-garde, in particular the interventions of Bauhaus wizard László Moholy-Nagy. In the work of both, mobile light beams never simply serve the purpose of illuminating existing sites or locations. Instead, they produce space as a dynamic field of movements, rhythms, and relations and in this they aspire to insert alternate worlds within the material confines of the given one. Unlike works such as Moholy-Nagy's *Light-Space-Modulator* (1930), however, Kuball's luminous exploration of movement is no mere exercise in playing out different logics of abstraction against each other. Similar to much of his own earlier work, Kuball's *res·o·nant* seeks to dramatize our experience of space, uses light to recalibrate the

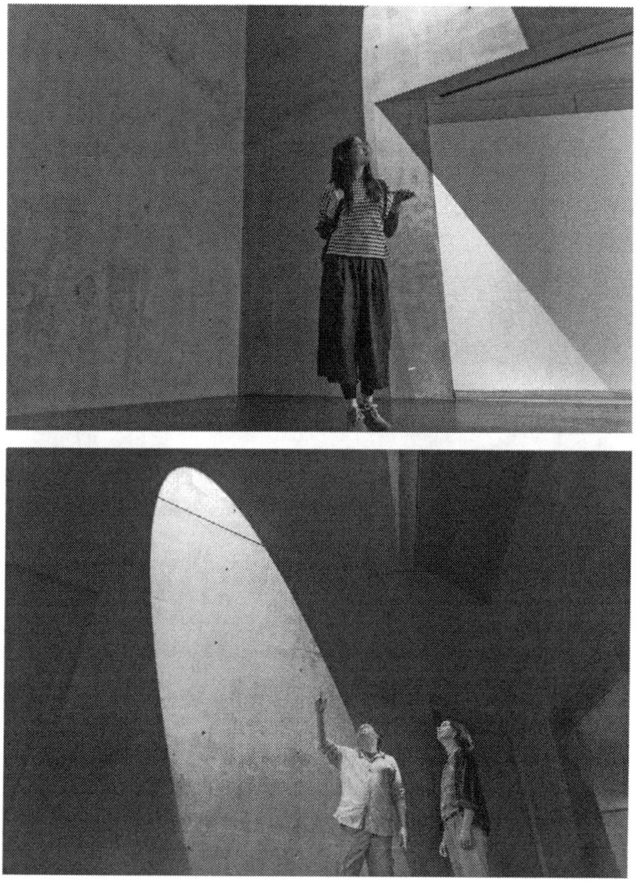

Figure 7.5 Mischa Kuball: *res·o·nant,* 2017–19.

relations of the public and private, and deliberately constructs radiant stages in order to address pressing moral and political questions.

Discussing several of Kuball's earlier works, Peter Weibel has noted:

> What resonates in Kant's famous phrase, "the starry heavens above me and the moral law within"—the association of light with morality, of aesthetics with ethics—is beautifully articulated in Kuball's light interventions. Kant and Kuball means: light can be used in the public space as a moral and ethical authority that provides a framework for social action as well as a recollection of it.[10]

[10] Peter Weibel, "The Politics of Light," in *Mischa Kuball: . . . in progress. Projekte | Projects 1980-2007,* ed. Florian Matzner (Ostfildern: Hatje Cantz, 2007), 270.

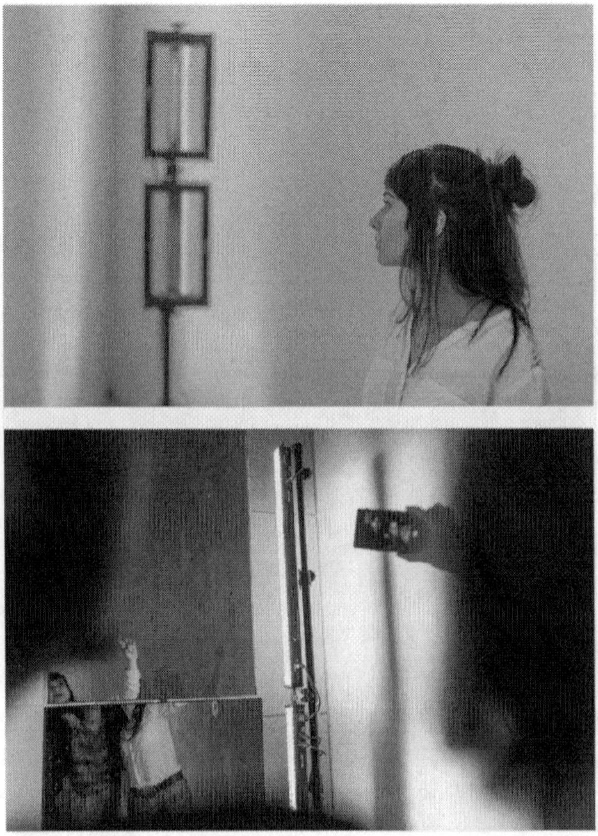

Figure 7.6 Mischa Kuball: *res·o·nant*, 2017–19.

Res·o·nant follows this impulse with its central ambition to probe and activate resonant interactions between architecture and skin, between Libeskind's stark concrete and the visitor's sensory perception. As Kuball's rotating shapes galvanize the visitors' attention amid the shafts' semi-darkness, *res·o·nant* asks the audience to reevaluate the architect's politics and ethics of space, that is, Libeskind's understanding of the building's voids as spaces sealed from the passage of time in order to signify the impossibility of ever re-membering the Holocaust rightly. In touching surfaces that symbolize trauma, loss, and absence, Kuball's shifting shapes relax the wall's solidity as much as they invite us to touch upon the void's walls at a distance. They serve as interfaces, and as such, they cannot but animate the public to rethink the stern ethics of Libeskind's design, the voids' inherent claim that no future could ever be able to resonate with the unlived lives left by the Holocaust.

But there is more. First, a rotating mirror construction in one of the two voids enables visitors either to come across unexpected reflections of their own presence or to actively conjure images of their own bodies—body images persistently pushing toward fragmentation or disintegration. Second, two stroboscopic lamps located opposite each other in the passageway connecting both shafts blast intermittent bursts of red light at the audience, assaulting the unprotected eye such that visitors cannot help but carry certain retinal afterimages with them as they enter the shafts themselves. Third, and most important for our context, 2 sets of speakers in each of the voids broadcast a looped series of 60-second sound clips—so-called skits—into the shafts, composed by more than 220 musicians and sound artists in response to an open call by Kuball and the museum's curator, Gregor Lersch. These recordings range from ambient drone sounds to acoustical experiments exploring resonances between different string instruments or between glass jars and birdsongs; from excerpts of Scriabin's works for piano to the sounds of barbed wire recorded at former South African prison camps; from the crackling of barely audible electric impulses to field recordings carried out in natural settings; from reverb and delay triggered by modular synthesizers to synesthetic experiments translating color perceptions into different musical patterns; from pleasing atmospheric sound carpets to notes that go right under the listener's skin; from samples of the pondering voice of German philosopher Alexander Kluge to the shrieking sounds of shattering glass, meant to recall the historical event of Crystal Night in November 1938. Some artists offered single entries, while others submitted a whole string of skits. Several contributors recycled old work, but most composed, recorded, or resampled new work. Most pieces were produced by individual authors, a few—such as the skits by the Berlin-based MonikaWerkstatt—resulted from the work of entire collectives. What all clips in spite of their vast differences share in common, however, is that all run for exactly sixty seconds while brief descriptions of the work flash up on tiny monitors in each of the shafts; and that each skit will play in one void while another clip will resound in a slightly staggered time window in the other.

Res·o·nant's spelling was drawn from the word's entry in *Merriam-Webster's Dictionary*, with the perforating dots indicating phonetic pauses separating different syllables from each other. *MW* defines "resonant" as "1: continuing to sound: ECHOING. 2a: capable of inducing resonance. 2b: relating to or exhibiting resonance. 3a: intensified and enriched by or as if by resonance // a *resonant* voice. 3b: marked by grandiloquence."[11] The sounds of Kuball's *res·o·nant* create far from predictable echoes between the visible and audible, architecture and skin, the shafts' vertical volumes and the bodies that navigate them. While the light projectors rotate according to expectable patterns, at once replicating and refracting the voids' footprints as luminous shapes, the installation's sounds cause audiences to pause mid-stride, listen into ever-shifting and at times quite demanding soundscapes, attune their movements to the sounds and beats, the drones and noises that reverberate through the voids. The title's dots thus signify at once rupture and connection: a coupling of the visual and the audible that, far from transforming the voids into sites of synthetic immersion, stresses disjuncture, difference, and disparity. Libeskind named his architectural design "Between the Lines" to emphasize the building's effort to represent what resists representation and display history as a process of unreconcilable trajectories, memories, and narratives. Similar to the installation's sights and sounds, the dots of *res·o·nant*, as they highlight the tensions between written and spoken language, push against any smooth integration of different media channels into what Wagnerian discourse once considered a total work of art. Haunted by the violence of the past, the present cannot but be experienced in a continual state of fragmentation. It resonates with the viewer's senses as something that, pace *MW*'s third definition of resonance, forbids any gestures of grandiloquence.

Since the early 1990s, Kuball's use of light has repeatedly transformed public space into a laboratory encouraging audiences to challenge the unease Germans continued to display when trying to address the past and the presence of Jews in Germany after the Holocaust. *Res·o·nant* recalls Kuball's earlier interventions, yet instead of primarily seeking to en-lighten the amnesia that had haunted Germany during the postwar period, the work for the Jewish Museum now leads the public straight into the house of memory and asks us to reassess what it means to remember, re-member, in the first place. The installation's stroboscopic lights and disassembling mirrors communicate in no uncertain terms that this process of re-membering the violence of the past can never achieve unity or closure: the dead are really dead, no act of recollection can ever fully right the wrongs of history. However, in his effort to transform Libeskind's voids such that architecture becomes skin and abstract concrete touches upon the visitor's sensory systems, Kuball arguably recalibrates the academic conceptualism, or, even better, the curious mix of deconstructive and quasi-expressionist stances,

[11] https://www.merriam-webster.com/dictionary/resonant.

Libeskind himself brought to bear on questions of memory when designing the building. Rather than conceptually seal off in time and space the memory of unlived lives, *res·o·nant* stresses the fact that voids of history—the silence of the murdered—require our ongoing engagement, our ongoing efforts to re-member them, in order to be recognized as such. The sounds of the living might echo like raindrops in the wells of this silence, yet it is only through our continued efforts to elicit such echoes and affects that we can formulate an ethics of memory—a politics of memory that re-members and thereby resituates the claims of history's victims as we move into the future. Affect, write Melissa Gregg and Gregory Seigworth, positions the body "as much outside itself as in itself—webbed in its relations—until eventually such firm distinctions cease to matter."[12] Kuball's *res·o·nant* insists that we need to be both outside and inside the body in order to be able to carry memories of historical trauma, of loss, absence, and silence, forward and allow the voids of the past touch upon future presents.

Cruel Optimism

John Cage famously suggested that, all things told, there is no such thing as silence. When he visited an anechoic chamber at Harvard University in 1951, he expected the room to swallow all sounds, echoes, and reverberations only to be struck, much to his surprise, by the buzzing of his nervous system and his blood circulation. Many of Cage's later compositions would revolve around ideas of silence or around experiences of whatever noise might be audible during moments initially perceived as silent. No acoustic void, for Cage, was vacuous enough not to produce sound, to be sounded out and listened to. No absence of sound was silent enough not to be probed for its echoes and reverberations or to be embraced as a resounding well of compositional practice. "Until I die," Cage wrote in late 1957, "there will be sounds. And they will continue following my death. One need not fear about the future of music."[13]

It would be cynical to deny that traumatic histories leave different silences in their wake: the silence of those whose lives were ended or brutally rerouted, the silence of the ones shocked about and mourning the loss of lovers, family,

12 Gregory J. Seigworth and Melissa Gregg, "An Inventory of Shimmers," in *The Affect Theory Reader*, eds. Melissa Gregg and Gregory J. Seigworth (Durham: Duke University Press, 2010), 3.

13 John Cage, "Experimental Music," in *Silence: Lectures and Writings by John Cage* (Middletown, CT: Wesleyan University Press, 1961), 8.

friends, neighbors, and comrades, the silence of those unable to speak about or remember what they have witnessed, the silence of perpetrators eager to forget their complicity. We have come to think of trauma as an event so discontinuous in nature that it prohibits our ability to remember and recall. Traumatized subjects, it is argued, are halfway stuck in the past, unable to experience time as a medium of change, transformation, and development. They carry, it has been added, the past as a silent, albeit strangely detached, burden through their lives and cannot but fail to hear any echoes and connections between what was and what is to come. Traumatic histories, it is concluded, place death at the center of the living. They partially deanimate their survivors. They situate anechoic chambers amid our systems of perception seemingly absorbing whatever may reverberate and could refract the spell the past may hold over the present.

It would be cynical to deny the silences caused by violence and trauma, yet as Cage reminded us in 1957 no silence is ever as silent, no void as void, as it appears to be. There is always some sort of sound, some sort of echo, however challenging it might be to identify its origin and location. Moreover, as Cage's integration of perceived noise into the repertoires of music suggested, whatever we call the normal or ordinary is far less normal and ordinary than we want it to be. Rigid binaries juxtaposing noise and music, or sound and silence, are as misleading as clinical assumptions pitting the discontinuities of traumatic histories against the presumed continuities of ordinary life. Similar to how music entails noise, and sound reverberates through what may initially strike listeners as the dark hole of silence, so it is necessary to think of the everyday, not as a radical opposite of the traumatic, but—in Lauren Berlant's words—as a "zone of convergence of many histories, where people manage the incoherence of lives that proceed in the face of threats to the good life they imagine. Catastrophic forces take shape in this zone and become events within history as it is lived."[14] Noise resides both at the edges and amid music, and so does the everyday already embed the incoherence of trauma and will therefore never simply be able to return to a state of ideal equilibrium should the impact of catastrophic histories finally recede. As sound can echo through and animate the wells of silence, so the everyday can absorb and reanimate the dead matter of traumatic experiences more effectively than cultural theory often assumes.

Kjartansson's *The Visitors* tests the viability of sound amid the silence following a painful breakup. Margolles's *La Búsqueda* embraces resonant vibrations to

[14] Lauren Berlant, *Cruel Optimism* (Durham: Duke University Press, 2011), 10.

move the memory of the missing from past voids into future presents. In Kuball's *res·o·nant*, the interplay of light and sound asks viewers to wear recollections of traumatic histories on their bodies—as second skins, as prosthetic limbs. Though the objects of remembrance in all three pieces are substantially different, each of these installations folds material and metaphorical aspects of resonance into dynamic logics so as to ask audiences to attend to the echoes and reverberations that may (re)animate the voids of history and embed the silence of unlived lives in the present. Their point is not to fill what is absent with new clamor and thereby deliver us once and for all from the past. Instead, it is to approach a present of traumatic arrest and animated suspension with what Berlant calls cruel optimism, that is, to develop different attachments to what is precarious, void, or causes pain to renegotiate some sense of ordinariness. Cruel optimism identifies impasses of time as media to retune our entanglements with the world and attend to what we may otherwise displace as silent, void, or unspeakable. No attachment can do without the promise to make something endure, survive, and flourish. Even at its most temporary, it articulates desires to stretch the present into the future and protect the past from oblivion.

What Kjartansson, Margolles, and Kuball share is their effort to offer aesthetic training grounds for cruel optimists. All three resort to the momentum of resonant vibrations to explore and, in the form of aesthetic promises, commit to what should survive and endure, including the memory of what has been silenced. The dead and the vanished need the living as resonant hosts of their memory. The living need to attach to the absent and departed if they hope to make good on claims of lives suspended in mid-stride. As suggested by the works of our three artists, cruel optimists concede that there is no absolute silence, no absolute void, no absence of echoes and reverberations as long as we simply train our ears and skins to listen closely. Well aware that the ordinary and the traumatic are closer than we often want to believe, they insist that in order to honor the memory of unlived lives and the agony of interrupted histories, we must—paradoxically indeed—embrace the vibrancies of the living and continually contemplate the extent to which we are able and willing to relegate death elsewhere. But . . .

. . . is Death Really Elsewhere?

Seven screens, installed in a circle facing inward (Figure 7.7). Each equipped with three speakers to propagate moving sound most dynamically. Their respective

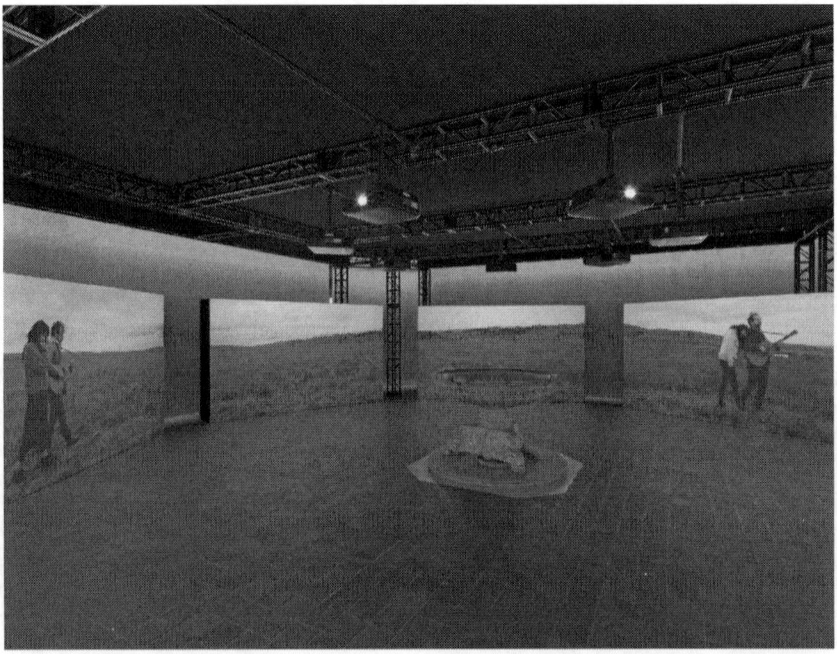

Figure 7.7 Ragnar Kjartansson: *Death Is Elsewhere,* 2019. Installation view at the Metropolitan Museum of Art.

images were initially recorded by seven cameras assembled in a circle as well, their lenses facing outward, such that what you see on the rotunda of screens closely matches what you would see when standing (and turning your gaze) at the site of recording. Taken together, the screens offer a striking panorama of the Icelandic countryside. Green grasses in the foreground, some gentle rolling hills in the middle, the dramatic sight of dark volcanic rock and lava moraines in extended areas of the background. The light is diffuse, giving no indication of daytime or cardinal directions. It blankets the viewer's perception, flattens out rough edges, provides little orientation, yet in doing so also comforts the eye, invites the gaze to settle in and take its time to peruse the landscape. As viewers of the installation, we are asked to stand inside the circle of screens, take on the position once occupied by the seven cameras. We are asked to immerse ourselves into this distant landscape, allow the installation's screen to transport our entire sensorium elsewhere even though our feet remain firmly placed on the ground. And relate to the movements on screen as if they are right in front of and around our eyes and ears.

The film that unfolds with a looped running time of seventy-seven minutes is as circular as the physical perimeters of the installation. We see two sets of identical twins walking in circles around us over and over again, each paired with a sibling of the other pair, each pair pretty much holding their position on opposite sides of both the real and the virtual circle (Figure 7.8). Their ceaseless steps have stomped the grass on the ground into a clear path, reminiscent of the lines Richard Long once left walking back and forth in the English countryside, yet at the same time humorously bending Long's play with ideas of passage and traversal into visions of repetition and circuitousness. Both pairs of walkers jointly sing a song together, the male members of each pair strumming a guitar as well. At times, they face their path, at times they look at each other, at times they seem to face us (or face the couple that at this point will be in our back). Their walking is as slow as their singing. As in so many of Kjartansson's previous works, not much seems to happen over the video's entire duration, thus ironically creating conditions under which more seems to happen and resonate with the viewer than anyone can process within one single viewing.

Two of the four performers in *Death Is Elsewhere* (2019) are well known from *The Visitors*: Gyða and Kristin Anna Valtýsdóttir. The other two performers are Aaron and Bryce Dessner, members of The National. They collaborated with Kjartansson in 2013 in the six-hour performance and recording of "Sorrow," a three-minute song repeated over and over again and consequently released as *A Lot of Sorrow*. The general conditions of filming the circular performances

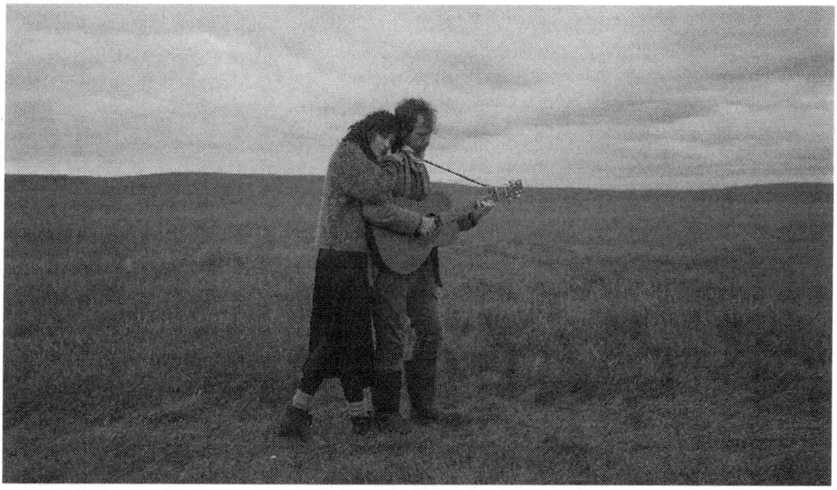

Figure 7.8 Ragnar Kjartansson: *Death Is Elsewhere*, 2019.

of two pairs of musician twins for the recording of *Death Is Elsewhere* was as difficult as it was to coordinate the schedules of all four of them. Certain segments of the landscape may look idyllic, but the lava landscapes that loom in the background are part of Iceland's forbidding Laki region, a volcanic fissure whose eruption in 1783 caused catastrophic famines across Europe and is said to be partially responsible for the social upheavals that resulted in the French Revolution.[15] Moreover, the recording of *Death Is Elsewhere* took place at around 2:00 a.m. close to summer solstice—a time that at a latitude of 64° N knows of neither darkness nor night and that provides perceptual conditions that may strike people with either exhilaration or insomniac despair. Or both at once.

The installation's title borrows from Alexander Dumbadzse's 2013 book on Dutch artist Bas Jan Adler, whose desire for the extraordinary led him to get lost at sea in his 1975 effort to cross the Atlantic in a thirteen-foot sailboat.[16] Even more minimalist than the lyrics of *The Visitors*, the repetitive stanzas of *Death Is Elsewhere* sing of love and death, of love stemming against the overwhelming forces of death and darkness, of love's emotive movements and streams keeping the stasis of death at bay: "In the dark / In the dark / My love / My love / By the stream / By the stream / My love / My love / Death is elsewhere." To be sure, unlike Adler, our twin bards will not vanish from life and go entirely absent. One of the two couples disappears from sight between the space of two screens for one entire round about thirty minutes into the video's 77-minute loop; and they all briefly fade from view toward the very end only to reappear as mirrored pairs as mysteriously on screen at the beginning of the next loop as they departed from the previous. As Gregory Volk writes: "It's hypnotic to watch them move from screen to screen: troubadours on the loose, minstrel poets wandering through the landscape, couples in love strolling through a meadow. The two couples become mirror images of each other; that they are two sets of identical twins makes things

[15] Here's how British naturalist Gilbert White described the outbreak's effect in 1783:

> The sun, at noon, looked as blank as a clouded moon, and shed a rust-coloured ferruginous light on the ground, and floors of rooms; but was particularly lurid and blood-coloured at rising and setting. At the same time the heat was so intense that butchers' meat could hardly be eaten on the day after it was killed; and the flies swarmed so in the lanes and hedges that they rendered the horses half frantic . . . the country people began to look with a superstitious awe, at the red, louring aspect of the sun. Qtd. In Richard Proctor, "The Climate of Great Britain," in *Intellectual Observer: Review of Natural History* (London: Groombridge and Sons, 1867), XI: 127.

[16] Alexander Dumbadzse, *Bas Jan Adler: Death Is Elsewhere* (Chicago: University of Chicago Press, 2013).

even more mirrored."[17] And yet, with the ominous sight of the Laki fissure at the horizon and midnight sun eventually to be eclipsed by winter's coming, death and darkness hover at the edges and beyond the path of our double twins, similar to how the ocean swallowed Adler's Nietzschean rhetoric of dangerous living. Death is elsewhere. But only barely. And only as long as we keep singing and dancing, watching and listening. Death is elsewhere. But only because art's rhythms and vibrations seek to push it there, ask death to do its work somewhere else, so that life, love, and meaning may temporarily reside here.

It is tempting to consider *Death Is Elsewhere* as a companion piece to *The Visitors*. The latter praises the tissues and networks of human resonance, however delicate they may be. It explores how sound emerges from silence in order to bar silence's lien on the living. The language of mirroring, of immersive enclosure, that structures *Death Is Elsewhere* amplifies and rechannels this. While the natural setting to some degree looks enchanting, the round's persistence leaves little doubt that nothing we experience as landscape is ever innocent. It requires human intervention, projection, and mediation to be defined as such and it may, therefore, be haunted by ghosts from the past that continue to drift through the present. What we understand as landscape here, within and beyond the perimeters of the round, is no more and no less than a screen, a panorama, a medium of human meaning-seeking and meaning-making. It embodies what Eyal Weizman, in a very different context indeed, calls elastic geography: the product of human actions to challenge, transform, and appropriate the land according to their needs, agendas, and desires,[18] the result of our efforts to control, exploit, find beauty in, or resonate with our surrounds. What is light and delightful, amiable and animate, is thus not merely borrowed from darkness and death; it in fact produces the very darkness and death, the indifference, silence, and violence that presumably exists beyond the elastic horizons of the human.

Rather than explore the voids painful histories and traumas have left in the fabric of time, *Death Is Elsewhere* contemplates models of how to live and be alive in face of the Anthropocene's catastrophic futures. The looming presence of the Laki fissure in the background serves as a stark reminder that the local, in all its tenderness and intimacy, cannot be neatly separated from the global, and that the human and the nonhuman are intertwined: nature and history effect each other across the planet's expanses in much more complicated ways than modern

[17] Gregory Volk, "Ragnar Kjartansson's Panorama of Love and Death," *Hyperallergic*, June 15, 2019, https://hyperallergic.com/505314/ragnar-kjartanssons-panorama-of-love-and-death/.

[18] Eyal Weizman, *Hollow Land: Israel's Architecture of Occupation* (London: Verso, 2007), 6–7.

textbooks of human progress have suggested. Everything is somehow connected, in particular in an age in which the human footprint has changed geological and meteorological conditions to such an extent that they now come to inflict traumatic catastrophes on us. Kjartansson's rotunda of screens does not merely represent nature; it is part of what now counts as nature. Yet this nature is no longer something we may enjoy freely because we have learned how to control or comprehend it. It no longer offers landscapes we traverse or celebrate because we know how to keep them at a distance. The twins' round allegorizes the present's immersive media landscapes as much as Kjartansson's circle of screens—their ecology of plastic, silicon, glass, and electricity—finds it analogue in the twins' dance and song. As we follow their tender movements, we cannot but come to understand that our new nature now entangles, encircles, and haunts us—eventually returning to us what we administered to it in our obstinate efforts to define what it means to be human. Death is elsewhere. And it is right here.

We need the death and its silence to be elsewhere to allow humans to reanimate their diamond hearts, hold on to the claims of unlived lives, and resonate with the voices of those that have been silenced in violent histories. But make no mistake: whatever urges us to relocate death elsewhere has long since invited death to return through the backdoor as well. Just as we may think that our sounds should host the memories of the vanished and departed, so we must recognize that our planet has lost its patience with our own volatile presence, that we, in overstepping the Holocene's hospitality, have created a world inhospitable to the human. *Death Is Elsewhere* asks us to re-envision earth's deep time and humanity's short-lived history, the histories of love and death, of trauma and healing, of past pain and present beauty, from the standpoint that we all are no more but mere visitors on this planet. In the larger scheme of things, our sounds, voices, and echoes will soon be void and silent (again). The planet as a whole will endure our self-destruction, similar to how it has endured past extinctions and has responded by giving birth to new and vibrant forms of life. There will be sound following our demise. We need not fear that our self-extinction will result in utter silence and the absence of echoes, resonances, even music. What's left for us, now and here, and with some sense of cruel optimism, is to resonate with the nature we have created; entangle our lives in networks of mutual relations; learn how to move, think, and act together with what is not us and build more accommodating homes than in the past; better understand temporalities whose vibrations precede and will succeed us; and use our remaining time to make good on our failure to live up to the unwritten laws of our planet's hospitality.

8

Telepathic Variations

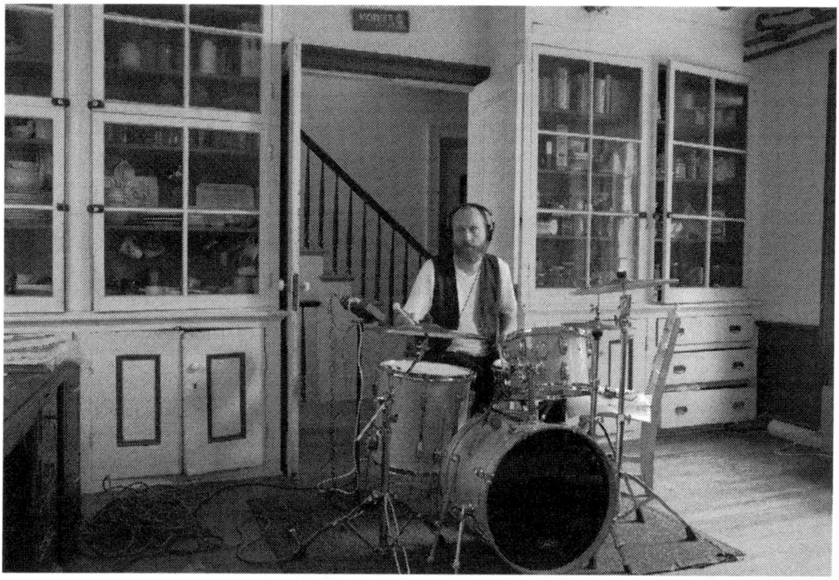

Figure 8.1 Ragnar Kjartansson: *The Visitors*, 2012 (Þorvaldur Gröndal). See also Plate 10.

Spooky Actions at a Distance

Eight musicians using headphones as a sign of both controlled deafness and intensified listening. They protect ear and body from their most immediate physical surrounds so as to enter an interior space of shared vibrations. A space that entangles sounds, emotions, and subjects precisely because it separates and reconfigures the relation of world and perception, mind and matter. A place that touches upon other places however distant they may be. Spooky action at a distance.

Since the introduction of the Sony Walkman in 1979, earphone listening in public has often been construed as a strategy to shut out the world in all its messiness in order to inhabit internal rooms of one's own.[1] David Toop reminds us that the compartmentalization of the sensible achieved through the wearing of earphones is much more nuanced:

> Lightweight earbuds allow ambient sound to mix with the listener's music, whereas the increasingly popular noise cancelling headphones create a relatively autonomous space that isolates the listener, not just from external sounds but from a sense of full presence within the environment. Ears enclosed by headphones, the body relinquishes the feedback function of its own sound. Contact between the body and its surroundings is so reduced as to be eerie and dreamlike. Instead, the body expands inwards by extending the perceived space of the mind. Music that can be recalled approximately through memory is suddenly reproduced in all its detail, apparently inside the head of the listener.[2]

Toop's observations are useful to think about the role of earphones in *The Visitors*. While the wearing of headphones is of course standard practice in any modern-day recording studio, their function in *The Visitors* is not about communicating with a distant recording engineer. Their principal job instead is to tune and retune the order of existing feedback loops. They move the audible world inside the head so as to expand awareness and consciousness outward, to get and be in touch with what transcends the visible. Worn by most of the players throughout the performance, they unlock networked spaces of mutual relations, of attachment and entanglement, as eerie and dreamlike as affective. As these headphones, in modifying the relations of materiality and consciousness, allow

1 For a compelling cultural history of the Walkman and its influence on public space, see Rebecca Tuhus-Dubrow, *Personal Stereo* (New York: Bloomsbury, 2017).

2 Toop, *Sinister Resonance*, 44.

our eight musicians to touch upon each other's sounds at a distance, they index no less than modern media culture's repeated toying with the idea of telepathy, that is, forms of communication that somehow bypass the "normal" operations of the senses, yet nevertheless trigger profound impressions in those able to receive distant signals.

First used as a term in 1882 by Frederic W. H. Myers, the modern preoccupation with telepathy—distant affliction, the transmission of affect and passion ("pathos") in the absence of obvious channels of communication—coincided with the rise of acoustic media such as the wireless, the gramophone, and telephony. As these new media all seemed to confront their users with ghost-like voices from locations distant in space or time, discourses on telepathy around 1900 aspired to recharge the perceived soullessness of modern industrial culture with new spiritual depths, meanings, and enchantments. In the realm of modernist art, telepathy's most iconic champion was Wassily Kandinsky, who rethought aesthetic communication as a telepathic process of emotional and spiritual transference, of creating affective correspondences between artists and viewers at a physical distance.[3] Art at its best, according to Kandinksy, enabled resonant interactions between body and soul. As it transported feelings through often invisible channels and, true to Kandinsky's well-known synestheticism, interwove different registers of sensory perception, it emancipated viewers from the crude materialism of industrial culture. Like the telephone, the wireless, and the spiritualist who animated the dead for the living, telepathic art built two-way bridges between matter and mind so as to reconstruct our access to the wondrous vibrations of the universe.

It is easy to brush away the association of art and telepathy as both a mystical and deeply paradoxical escape hatch from the dreary spaceship of modernity. Mystical because it aspired to roll back the presumed contamination of the numinous at the hands of modern technological innovations, scientific reason, and mass society. Paradoxical because art's desire to serve as a medium of the spiritual, the distant, and the resonant echoed the uncanny breakthroughs of modern media of telecommunication, yet at the same time sought to obscure any possible affinity between mediums and media. But even in Kandinsky's

[3] "The inner element," Kandinsky wrote in 1913, is the emotion in the soul of the artist. This emotion is capable of calling forth what is, essentially, a corresponding emotion in the soul of the observer. As long as the soul is joined to the body, it can as a rule only receive vibrations via the medium of the feelings. Feelings are therefore a bridge from the nonmaterial to the material (in the case of the artist) and from the material to the nonmaterial (in the case of the observer). Wassily Kandinsky, "Painting as Pure Art," in *Complete Writings on Art*, eds. Kenneth C. Lindsay and Peter Vergo, vol. 1 (New York: Da Capo, 2004), 349.

case, things were already more complicated than this. For Kandinsky's vision to communicate pathos at a distance relied on (theosophical) understandings that questioned any ontological binaries of distance and proximity, spirit and matter, reason and resonance in the first place. It mapped the world as a cosmos of vibratory energies that manifested themselves in myriads of ways within and beyond the perceptual range of the human senses. As mysterious and un-enlightened as this may have sounded to many of his contemporaries, Kandinsky's efforts to rediscover the spiritual qualities of vibrating matter curiously paralleled the newest insights of modern science and how twentieth-century nuclear physics was about to change Newtonian conceptions of materiality. In a famous presentation on Kandinsky's work delivered in April 1917, DADA poet Hugo Ball in fact associated Kandinsky's forays into the telepathic with how recent scientific discoveries—"the electron theory"[4]—had displaced mechanistic visions of the world and created a universe in which everything once considered solid, including former dualisms of matter and consciousness, now came to melt into thin air. While Ball left little doubt about the fact that modern technology had created many unheard-of monsters, he simultaneously stressed the affinities between Kandinsky's view of art as a medium of telepathic vibration and modern science's explorations of the delicate vibrations of all matter. Rather than repudiate the critical insights of modern science, theosophy's understanding of matter as a host of resonant oscillations offered the very flipside of how advanced scientific reason now sought to explain the world's innermost building blocks.

In physics, the concept of resonance typically presupposes mechanistic approaches to matter. We think of resonance—one body's ability to make another vibrate—in terms of a linear chain of influences and movements that confirm the principal continuity of the physical world. God does not play dice in this conception of the world, nor does matter behave unpredictably. Its blueprint has no place for chance or probability, luck or the accidental. Though the mechanistic principles of Newtonian physics reliably explain most of what meets everyday eyes and senses, post-Eisensteinian science of course no longer endorses this blueprint in its entirety. Research on the vibrancy of atomic and subatomic particles upset the very frameworks that continue to present resonance as a phenomenon of mechanistic continuity. It makes us wonder whether matter has much more in common with

4 Hugo Ball, "Kandinsky," in *Flight Out of Time: A Dada Diary*, trans. Ann Raimes (New York: Viking, 1974), 224–5.

affect, materiality with the illogic of consciousness and volitional agency, than Newtonian physics wanted us to believe. It makes us speculate whether the telepathic is the common order of, rather than the weird exception to, the way in which media mediate matter and mediums have the power to call forth what is distant.

Contemporary sound art has certainly a thing or two to say about this, not only when it shows little respect for the difference between literal and metaphorical understandings of resonance, but also when it—as in the three cases I discuss in this chapter—explores resonant relationships between advanced science and art, physics and metaphysics, head on. The bohemian postures of Kjartansson's earphone artists, as they telepathically communicate with each other across the spaces of Rokeby Farm and with us as their listeners in different sites across the globe, might be as distant as possible from the image we love to circulate about nerdy scientists. It may turn out, however, that Kjartansson and friends, as they fall into what their song's lyrics describe as "feminine ways," know more than your ordinary physics textbook about matter and materiality, information and communication, and the often spooky paths in which objects of this world affect each other.

Telepathy's main pathways today rely less on Kandinsky's mysterious mediums of paint and air than on the invisible work of cables, miniaturized speakers, Wi-Fi networks, and Bluetooth connections, attaching listeners to sonic vibrations whose origins in space and time are almost always, according to the original Greek meaning of the word "tele," far off indeed. And yet, similar to how telepathic discourse a century ago raised tricky questions about the nature of matter and the invisible undersides of information to which post-Newtonian science offered compelling answers, earphone listeners such as Kjartansson and his friends today stir our curiosity about what resides behind the visible, our desire to understand the visible *and* invisible logics of matter and communication, of what makes matter matter and how such mattering touches upon and travels across different bodies, media, and mediums.

Their resonant sounds do not simply perform science in a different key. They instead present art as a central medium to approach, inhabit, and know the world scientifically today.

Alpha Wave Music

On May 5, 1965, Alvin Lucier sat down on a chair in the Rose Art Museum at Brandeis University, asked an assistant to attach three sensors to his forehead and temples, and then closed his eyes in front of an audience curious about what was to come (Figure 8.2). The electrodes were connected to a series of amplifiers, a band pass filter, a mixing board, and finally a set of sixteen different percussion instruments, which after some time of waiting eventually started to resound. Lucier's attire resembled that of a mid-century bureaucrat while the wiring of his head evoked memories of bad science-fiction films. Once they reverberated from the different instruments on stage, however, the sounds of thumping and drumming had something mysterious, something magic and wondrous that was quite at odds with the sober technical setups on view. Recalling the laboratory of mad scientist Rotwang in Fritz Lang's *Metropolis* (1927), Lucier's futuristic configurations offered audiences very little to understand how one piece of machinery impacted the other. And yet, whatever music Lucier's virtual orchestra produced to fill the auditorium with sounds resulted from a very deliberate calculus, interfacing advanced scientific tools with seasoned

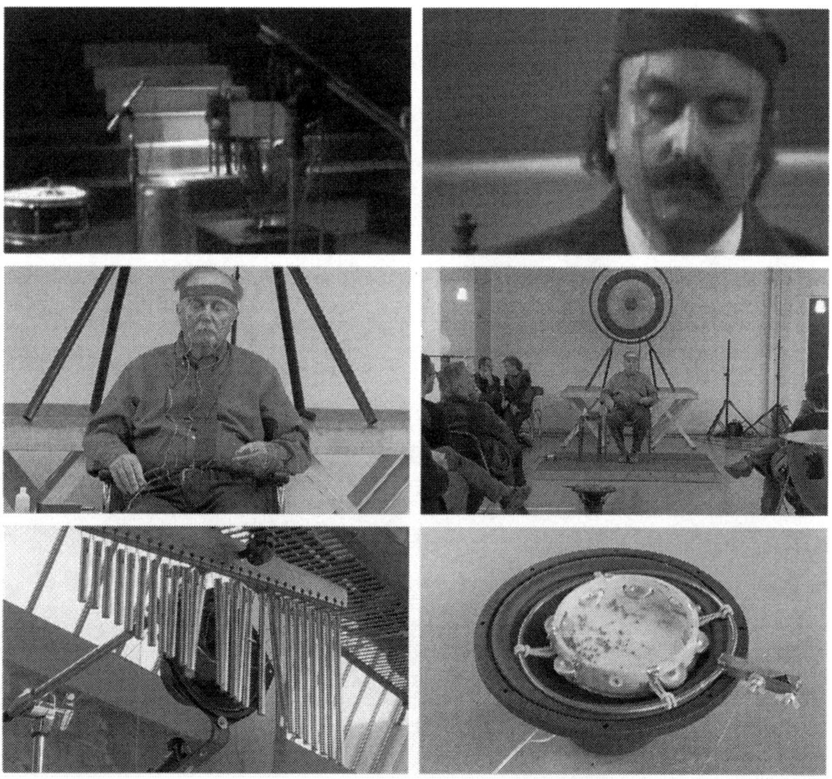

Figure 8.2 Alvin Lucier: *Music for Solo Performer*, 1965, 2010.

instruments of artistic expression. The score for *Music for Solo Performer*, written a while after the actual performance, described the transformation of brainwaves into musical intensities as follows:

> Place an EEG scalp electrode on each hemisphere of the occipital, frontal, or other appropriate region of the performer's head. Attach a reference electrode to an ear, finger, or other location suitable for cutting down electrical noise. Route the signal through an appropriate amplifier and mixer to any number of amplifiers and loudspeakers directly coupled to percussion instruments, including large gongs, cymbals, tympani, metal ashcans, cardboard boxes, bass and snare drums (small loudspeakers face down on them), and to switches, sensitive to alpha, which activate one or more tape recorders upon which are stored pre-recorded, sped-up alpha. Set free and block alpha in bursts and phrases of any length, the sounds of which, as they emanate from the loudspeakers, cause the percussion instruments to vibrate sympathetically. An assistant may channel the signal to any or all of the loudspeakers in any combination at any volume, and, from time to time, engage the switches to the tape recorders. Performances may be of any length.[5]

Alpha waves are neural oscillations in the range of 8–12 Hz. Even if we had the ability to listen straight to the brain's vibrations, they remain well below the threshold of the audible. They were traditionally believed to represent states of utter inactivity in the visual cortex. More recently, neuroscientists have argued that they play more complex functions, including coordinating the network of certain brain activities. What remains uncontested, however, is the fact that alpha waves are critically involved in sleep-wake cycles and peak when research subjects are in states of relaxation, their eyes wide shut, their bodies neither tired nor asleep. Their hub is the occipital lobe, the visual processing center of the cerebral cortex, so any effort to actively produce bursts of alpha waves needs to begin with the paradoxical effort to halt one's drive to visualize, to visually imagine spatial relationships. To want alpha waves you have to let go of really wanting anything. To set them free from your brain you need your brain to restrain its operations, urging it to fall into a state of vibrant stillness.

Lucier's *Music for Solo Performer* aspired to face this paradox. "One is supposed to get into a relaxed, meditative state to unblock it [alpha]," Lucier recalled later,

[5] Alvin Lucier, "Music for Solo Performer (1965) for Enormously Amplified Brain Waves and Percussion, Score," in *Chambers: Scores by Alvin Lucier. Interviews with the Composer by Douglas Simon* (Middletown, CT: Wesleyan University Press, 1980), 67.

"but I learned to do so with relative ease."[6] Sonifying the brain's unsound, eight home stereo amplifiers in the 1965 performance sent amped-up oscillations to sixteen different speakers whose vibrations in turn caused the chosen percussion instruments to vibrate themselves, to resonate. Initially planned to last ten minutes only, the first performance in the end took forty minutes, not least of all because Lucier required more than ten minutes in the public setting to produce bursts of alpha in the first place. In his extended discussion of Lucier's intervention and collaboration with neuroscientist Edmond Dewan at Brandeis, Douglas Kahn describes Lucier's performance as follows: "While Lucier attempted to refrain from mental activity, percussion sounds slowly started to fill the room, which were suddenly disrupted when he opened his eyes, engaged in mental exercise, or when his attention was drawn towards sounds from the audience."[7] There is no real need to point out, though, that the piece's title was rather misleading. Lucier's willful embrace of meditative non-visualization may have rattled traditional concepts of authorship and creative sovereignty. The title of his performance, however, covered up the crucial work of his assistant to channel oscillations to different resonant instruments and thus largely control the entire sound of the piece.[8] And yet, none of this should call into question the most compelling provocation of *Music for Solo Performer*: its effort to engage recent scientific and technological advances, not only in order to reframe our understanding of resonant interactions, but in so doing also to bring into play perceived tensions between the visible and the invisible, between the matter of the mind and the vibrant materiality of sound.

In the mid-nineteenth century, the work of Hermann von Helmholtz was critical in abandoning various tools of visualization in order to understand the work of audition. Instead of constructing objects that translated acoustic vibrations and the perception of sound into something observable with the eye, Helmholtz's effort was to focus on the ear itself and its neurological attachments as complex media of transmitting largely invisible vibrations.[9] In lockstep with the newest technological innovations of his times, Helmholtz eventually compared the body's and brain's networks of hearing and perception to the invisible work

6 Lucier, *Music 109*, 52.

7 Kahn, *Earth Sound Earth Signal*, 89. For more about the science behind Lucier's performance, see Bart Lutters and Peter J. Koehler, "Brainwaves in Concert: The 20th Century Sonification of the Electroencephalogram," *Brain*, 139.10 (October 2016): 2809–14.

8 Volker Straebel and Wilm Thoben, "Alvin Lucier's *Music for Solo Performer*: Experimental Music Beyond Sonification," *Organised Sound*, 19.1 (2014): 17–29.

9 Steege, *Helmholtz and the Modern Listener*, 44.

of telegraph wires, transporting electric impulses whose impact would greatly vary depending on what kind of objects they connected and set in motion. For Helmholtz, the nerves that enabled the brain to register images or sounds were no different from the wires allowing humans to exchange written messages in encoded forms across vast distances. In fact, human nervous systems and brains were best understood as electric power grids: networks of flows, transductions, bursts, and stimuli at once emulating and illuminating the rise of modern mass media and technologies of telecommunication. Everything depended on the invisible oscillations streaming through nerve fibers and connectors, yet these oscillations in themselves were agnostic to the sources that triggered them. Similar impulses could result either in a sound heard or in a smell smelled, each outcome simply depending on the brain's and body's wiring scheme.

Lucier's *Music for Solo Performer* simultaneously culminates and reverses the departures of nineteenth-century science. Its view of the human subject is that of a power plant generating formal energies, flows, and oscillations necessary to set things in motion and cause perception to resound along with them. Electrodes, wires, amplifiers, and speakers here transform the brain's and nervous system's energies into vibrations able to resonate with remarkably different percussion instruments. Like Helmholtz's fibers, they mediate echoes between highly incongruous entities and thereby entangle world and perception with each other. Contrary to Helmholtz, however, Lucier largely externalizes the very networked processes the German scientist hoped to locate within the human body. As they resonate with the impulses generated by the artist's alpha waves, Lucier's timpani, snare drums, and cardboard boxes now serve as subjects of perception, whereas the brain and our nerves, precisely in trying to abstain from activity, produce the very signals necessary to enable perceptual processes. World becomes brain, brain becomes world. Perception resides in things, while resonance appears emancipated from the fibers of human subjectivity.

Lucier's performative world of resonating alpha waves has neither place nor need for the work of earphones. As it positions the cerebral cortex as the source of sound and musical instruments as its vibrant listeners, it in fact does away with traditional paradigms that think of music as a linear chain of causes, effects, and affects traveling from human to human. Whatever is "solo" about *Music for Solo Performer* recasts music as something that could never be reserved for private and undisturbed acts of listening nor best enjoyed when shutting out the rest of the world. The vibrations of sound live everywhere, whereas hearing and listening are by no means confined to the organs of living beings. The musician's task, in all this, is simply to step back, let go, fall into what neuronal networks do best when left undisturbed. It is to serve as a host for viable flows of electric impulses and as a steward facilitating the journeys of waves and particles across different materialities. Science here teaches art that music owns many more lives and (inter)faces than musicians have ever assumed. Art in turn allows science and technology to realize that neither the production of knowledge nor the development of new tools need solely to serve the purpose of disenchantment, of defining the human subject as a sovereign master over the world and its contingencies.

With this, *Music for Solo Performer* ironically meets some of the same ambitions Ragnar Kjartansson and his collaborators brought almost fifty years later to their performance of *The Visitors*. Lucier's world of resonant alpha waves is one in which sound and music are far too expansive to be contained by devices such as headphones: music here in fact matters because the matter of music suspends traditional views of the listening subject, stresses structural reciprocities between acoustic media and the operations of the brain, and in this way disbands as illusion the very possibility of privatized or sovereign listening. The microphones, wires, and headphones of *The Visitors* are tasked to achieve something quite similar starting from the opposite side of the spectrum. They ask their users to throw away the ladder of privatized sound once they have climbed it only to discover that in the world of music, expansively understood, everything is connected indeed so that the diverse and solitary can resonate with each other across their very difference.

Quantum Solace

One artist, Philipp Lachenmann, takes us down into the underland of CERN and captures the performance of improvised Indian raga music in front of the DELPHI detector, one of four former sensors along the Large Electron-Positron Collider (LEP). The underground tubes of the LEP were decommissioned in the early 2000s to make space for the construction of the Large Hydron Collider (LHC); in 2012, this newer 27-kilometer circular accelerator played a critical role in confirming the existence of the Higgs boson, a particle so elementary in giving mass to all other particles that—much to the dislike of nuclear physicists—public media labeled it as the God particle. Though no longer used for scientific experimentation, the DELPHI detector has remained in place, its at once massive and intricate mesh of wires serving as a monument to the work nuclear physicists have been doing at CERN to offer, at heart, succinct answers to two questions: "What is the nature of our universe?" and "What is it made of?" Lachenmann's work suggests that, in order to approach possible and probable answers to these questions, we cannot afford to ignore the generative power of the aesthetic—the resonant voice of art.

Another artist, Carsten Nicolai, leads us into a high-tech laboratory of his own design. Two large mirrors, each mounted to an Archimedean body split in half, face each other in a long and narrow gallery space. Two laser beams travel back and forth between these two objects, each hitting a photocell detector attached to the opposite mirror whose activation—in what turns out to be a self-reproducing system—prompts an adjacent laser beamer to send a beam back in return. The lasers' parallel sharp green beams dominate the space of the installation, simultaneously interconnecting the divided solids and featuring their spatial distance. Laser beams travel at the speed of light, and yet what we encounter in Nicolai's installation appears at first to feature time at a standstill: light so steady, straight, and stable in nature that it assumes the qualities of a solid sculpture, a tangible object. Nicolai, however, wants to take his viewer elsewhere. Rather than showcase classical notions of timelessness and circular time in the garb of advanced technology, the apparent tranquility of the installation explores how modern physics' investigation of complementarity, quantum entanglement, and teleportation—what Einstein with profound reservations called spooky effects at a distance[10]—unsettles traditional understandings of time and space altogether and makes us think of the nature of matter as being reliant on processes of observation and, for the lack of better words, consciousness and perception.

We know little, writes Robert Macfarlane, about the world(s) that exist below our feet.[11] Language privileges experiences of ascent and uplift, whereas what reside underneath the earth's surface is associated with depression, catastrophe, and cataclysm. To look upward is to look forward and envision progress and transformation. To descend into the netherworld is to, at worst, go to hell, at best get entangled with deep, seemingly static nonhuman times. Lachenmann's *DELPHI Rationale* (Figure 8.3), first installed at the Schering Stiftung in Berlin in April 2018, carries us into the uncanny, the otherness, the sublimity of the

[10] Albert Einstein, Boris Podolsky, and Nathan Rosen, "Can Quantum-Mechanical Description of Physical Reality Be Considered Complete?" *The Physical Review*, 47.10 (1935): 777–80.

[11] Robert Macfarlane, *Underland: A Deep Time Journey* (New York: W. W. Norton & Company, 2019), 14.

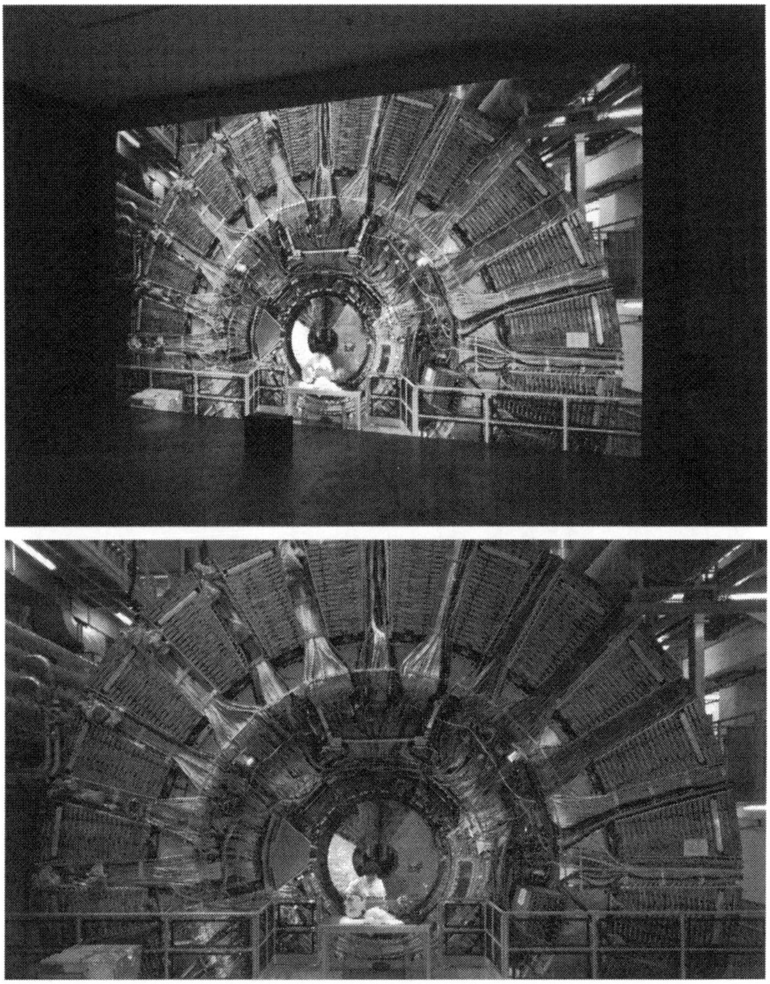

Figure 8.3 Philipp Lachenmann: *DELPHI Rationale*, 2018.

underland, a territory in which clocks tick differently than above ground. Most of us admittedly know as little about CERN's colliders, let alone exactly what they do to reveal the nature of the universe, than about the many sediments of nonhuman histories that reside underneath our feet. Both, in fact, in expanding, contracting, fraying, or simply ignoring the temporal orders humans cultivate to structure their existence, question what we might mean with knowledge in the first place. The deep time of caves, catacombs, and arctic moulins calls on us to recalibrate our knowing as much as the fast-paced spinning of charged particles below Swiss ground. They urge us to emancipate knowledge from the desire to "grasp" matter at hand—to probe forms of knowing that echo and resonate with what exceeds our abilities to categorize and reason.

Lachenmann's *DELPHI Rationale* is shot in one extended static take. The image initially shows us the monumental face of the DELPHI detector for about ten seconds before Indian musician Pandit Ranajit Sengupta enters the frame from the right, walks up to the center of both image and detector, sits down with his sarod, its many strings visually echoing the multitude of wires that make up the apparatus behind him, turns on a music player delivering atmospheric electronic sounds, begins to play his instrument for the next eight minutes and then eventually places the sarod on the ground, wraps a traditional robe around his shoulders, and exits the frame on the right—only to leave the viewer with the sight of the LEP for another three minutes in which the camera's continued stillness serves as a counterpoint to the playful flow of colors Sengupta's raga sounds have seemingly unleashed in and across the detector's weave of cables. The Sanskrit word "raga," as it has come to describe certain melodic forms of traditional Indian music that thrive on improvisational elements, means to tinge with color or dye. It describes musical practices that aspire to strongly affect their listeners by coloring their minds. And color is at the heart of *DELPHI Rationale* indeed, not only in terms of what we hear, but equally in terms of what we see, first with some sense of incredulity and wonder, then as the video proceeds with the eyes of spectators who understand special effects not simply as tricks fooling our senses but as central to the relation of humans and media, as technologies of humanization and techniques of humanizing machines.

Attuning CERN's high-tech scientific research environment to the sounds of Sengupta's sarod, the play of color behind the musician's back begins gently and first almost imperceptibly a few seconds into the performance. It gradually builds up in intensity and crescendos once Sengupta leaves the stage in front of the detector's eye, as if the initial impact of resonant vibrations now allowed the apparatus to live a life on its own. When not subjected to intricate processes of cinematic postproduction and digital color manipulation, DELPHI's filigree of wires is dominated by dirty oranges, blues, and greys. In Lachenmann's version, these cables pulsate in eight basic colors, all sampled from classic Technicolor films such as *The Garden of Allah* (1936), *The Wizard of Oz* (1939), and *Gone with the Wind* (1939), transposing the historical use of color in cinema as a vehicle of fantasy production, imaginary travel, and awe into a mesmerizing choreography of visual rhythms. Technology here approximates magic, scientific reason the oracular that echoes in the detector's acronym, yet not, as early twentieth-century critical theory would have argued, in order to displace critical reflexivity with mind-numbing phantasmagorias, but to illuminate the extent to which particle physics at once transcends and energizes the work of affects and the imagination.

Since its founding in 1954, CERN's scientists have illuminated the extent to which any possible answer to the question of the origin of matter requires us to rethink what traditional physics understood as matter—the ideas of continuity and contiguity, cause and effect, irreversibility and predictability we typically invest into our understanding of the world's material building blocks. A world in which the color of seemingly inanimate objects curiously echoes the vibrations of sounds, Lachenmann's underland defines a realm in which the rhythms of matter and mind touch upon each other, not because the artist hopes to move scientific reason over the rainbow, but because in his view Oz and CERN have more in common than either might initially be prepared to admit. Both color the mind and descend into alternate orders of time because in both the world of material facts cannot be separated from the work of perception. In both perception and its affects belong as much to humans as to superconducting magnets and electromagnetic calorimeters, tin cans and scarecrows. And in both matter cares for the movements and life of other matter in no way different from how human subjects may take care for and aspire to resonate with other human subjects.

While it is easy to dismiss Lachenmann's interweaving of science and fiction as speculative mysticism, it is worth to recall that its grounds have been mapped out by some of the most esteemed twentieth-century physicists themselves. For instance, Werner Heisenberg in 1927 held that it is impossible to measure both the velocity and the position of an object with exactness at the same time, not even in theory. Uncertainty about matter cannot but prevail; even the most advanced scientific gadget cannot eat and have the cake at once. Similarly, Niels Bohr in 1928 found that the most basic building blocks of matter can be seen and understood either as waves or particles. One view complements the other, yet it is impossible to observe, let alone calculate, both properties simultaneously. With Heisenberg and Bohr, modern physics dislodged the certainties science once had expressed about the order of matter. In their wake, the world no longer appears as a collection of facts, but of competing probabilities and projections. The properties of matter change according to observational setups. Nuclear particles attune their trajectories or positions to the standpoints of their observers; they in fact remember and even anticipate the movement of other particles and behave accordingly. They communicate across discontiguous spaces and discontinuous times, in patterns that cause scientists to see uncanny processes of time travel and teleportation at work, and that in fact cause some of the most unsettling insights of Einstein's thought to look as if they belonged to an orthodox and now untimely regime of atomistic determinism.

Sengupta's sarod and DELPHI's spectral color flows quite accurately describe a post-Heisenbergian world in which particles resonate with each other and their observers and adjust their movements to the oscillations of others, yet in so doing no longer follow or subtend any mechanistic understanding of resonance based on the idea of unmediated physical induction. Newtonian concepts of time as a linear unfolding of cause and effect, and of space as a realm in which rational minds encounter mindless matter—the fulcrums of how material sciences prior to modern nuclear physics approached the work of resonance—here no longer hold. Rather than ushering advanced science and technology over the rainbow, Lachenmann's *DELPHI Rationale* illustrates that they have come to inhabit Oz for some time now themselves, so much so that we have no real reason to hang on to the difference between seasoned understandings of resonance as either deterministically unambiguous or metaphorically and affectively fuzzy. The relation between different particles is not so different from the rather messy and often enigmatic one between musical sounds and their listeners. A musician able to touch bodies and minds at a distance knows as much about the nature of the universe as particle physicists. In Lachenmann's *DELPHI Rationale*, the ability of raga to color the mind expresses art's quantum solace. It illuminates the degree to which science itself has long abandoned C. P. Snow's famous postwar talk of the existence of two incompatible cultures in the allegedly fully disenchanted West.

Distant Entanglements

First installed in 2018 at the Berlinischen Galerie in Berlin, Carsten Nicolai's *tele* radicalizes Lachenmann's intervention (Figure 8.4). Its sense of cold, geometric abstraction may at first annul *DELPHI Rationale*'s play with the wondrous and enchanting; there is little room in face of Nicolai's symmetric laser beams and Archimedean bodies for the charms of (Techni)color. As we have seen when discussing different versions of *wellenwanne* earlier already, Nicolai's work often hovers in the space between art and science. His installations resemble experimental setups; they offer laboratories to probe the dynamic interaction of physical matter, visible or not, with the audience's senses. In *tele* Nicolai concerns himself with particle phenomena that defy any human scale and hence escape

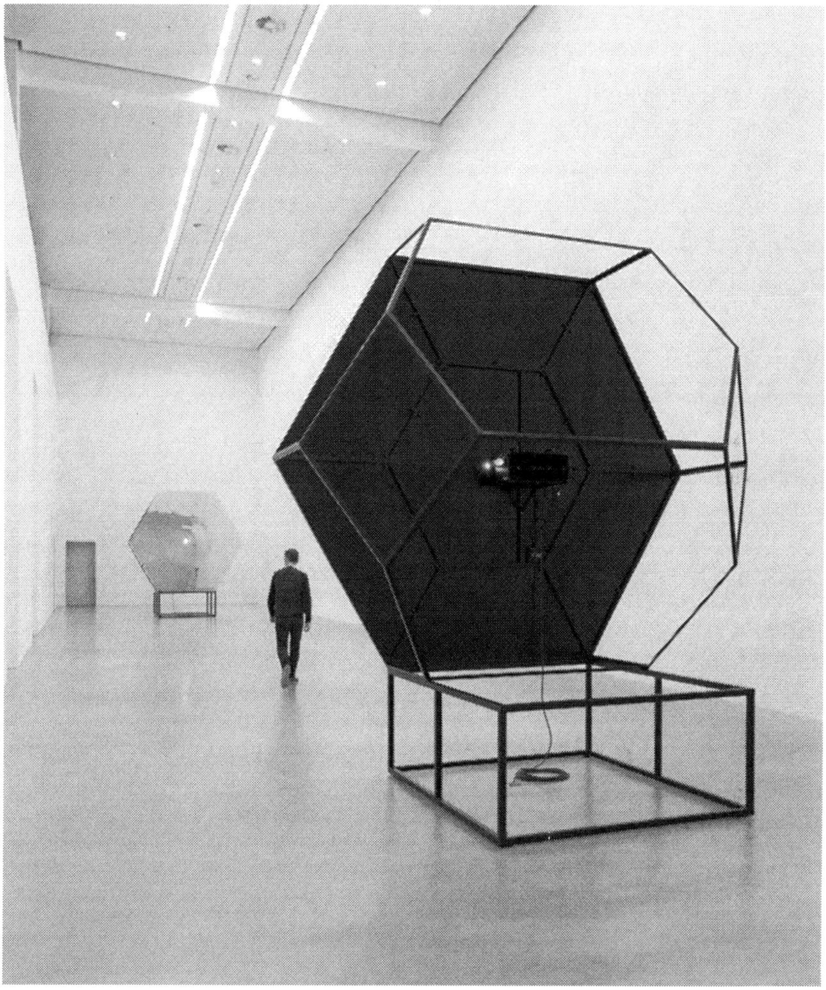

Figure 8.4 Carsten Nicolai: *tele*, 2018.

visualization, sonification, and sensory perception: quantum teleportation and entanglement, that is, processes in which atoms communicate information about their location and movement to other particles in defiance of Newtonian or even Einsteinian understandings of spacetime.

Most famously conceptualized by Austrian physicist Anton Zeitlinger,[12] theories of quantum teleportation and entanglement elucidate how measurements

[12] Anton Zeitlinger, *Dance of the Photons: From Einstein to Quantum Teleportation* (New York: Farrar, Straus and Giroux, 2010).

about certain particles in one specific location in spacetime can affect particle behavior and measurement at different points, even prior to the time it would take for anything to travel at the speed of light. Teleportation describes the movement of information, not of matter itself: Trekkies therefore find little in Zeitlinger's work to corroborate Scotty's ability to beam bodies across space. And yet, with its stress on informational exchanges that unsettle anything we know about the materiality of media connecting sender and receiver, theories of quantum teleportation conceptualize a universe in which not only everything somehow resonates with everything else, but also concepts of irreversible chronological time, and of the velocity of light as the speed limit to all thinkable movement, no longer accurately describe the (dis)order to things.

The self-enclosed system of Nicolai's *tele* allegorizes what eludes both allegorization and embodiment. Matter, in the form of Nicolai's Archimedean solids, here communicates across spatial distances at speeds so fast in nature that time appears at a standstill. As the two laser beams—at once wave and particle, light and electromagnetic radiation—continuously trigger each other's existence and thereby mark time as recursive rather than linear, they model the entanglement of matter across distant spaces. Though the Archimedean solids in this setup elude movement and oscillation, Nicolai encourages us to think of their interaction, their transaction of information, as a mode of affective resonance, of tele-pathos. One echoes the location of the other. Each is aware of the other and in fact requires the other's awareness and perception to perform its own.

What is perhaps most striking about Nicolai's setup is the fact that, in staging the resonance of matter, it leaves little for both artist and audience to do once the process is set into motion. *Tele* at first encounters the viewer as utterly predictable, so much so in fact that the installation seems to shut out the audience: particles here teleport and entangle in cold defiance of human observation. And yet, in tapping into the weird and recursive temporalities of quantum mechanics, *tele*'s self-contained system of nonhuman resonance of course hints at the opposite. It unsettles the very orders of (human) temporal perception that allow us to think of something as predictably unfolding in linear time to begin with. Time, as we know it (at least ever since Kant), is only for us; it is what we bring to the world of matter, to understand and grasp it, to order and lay our hands on it. Time is our interpretation of time, a portal we chisel into the world to make it pliable. We call something boring, iterative, predictable, surprising, or new only because time as a category of our understanding presents it as such to us. *Tele*'s staging of atomic resonance indicates the possibility of alternate temporalities, the existence of a world that is not for us: a material world that—like Kant's thing-in-itself—exceeds or defies our understanding, and that we may experience as predictable only because we are habituated to expect things to move in time and space on our own familiar terms. Everything in *tele* appears under control and hence rather static precisely because in the world of telepathic particles nothing really confirms to our ideas of control.

It is in this, then, that the aloofness of Nicolai's *tele* curiously approximates the dazzling magic of Lachemann's *DELPHI Rationale*. While the latter approaches the world of nonhuman objects as if it was utterly charged with subjectivity, the former stages the real as being incompatible with human understandings of the subjective. Both, however, not only find processes of telepathic resonance at work in strange places, whether they make sound or not. Both in doing so question convenient dualisms of subject and object that reflect the modern understanding of science and technology as tools enabling humans to dominate, control, or re-engineer the vagaries of so-called nature. Whether it is for us to grasp or not, both Lachenmann's and Nicolai's world of matter is one endowed with perception and pathos, memory and anticipation—and hence perhaps even consciousness. It is vibrant to its core, perfectly able to host processes of telepathic communication and resonant infection, even though such vibrancy and pathos tends to escape our human perception and understanding. As a privileged medium to explore the strange and incommensurable, art in both Lachenmann's and Nicolai's work can engage and resonate with advanced nuclear physics because the vibrant underland of atomic and subatomic particles is as much driven by uncertainty and empathy as the world of aesthetic objects.

Ragnarök Revisited

Ragnar Kjartansson is no scientist. Not even close. Unlike the work of Lucier, Lachenmann, and Nicolai, Kjartansson makes no explicit effort to emulate in art processes of scientific experimentation, let alone provide metaphors and aesthetic laboratories to recalibrate the role of scientific reason in our contemporary world. His work neither features the detached sobriety of Nicolai nor Lachenmann's spiritual focus. It has little in common with Lucier's ambition to tap into advanced science to remake traditional concepts of artistic authorship, nor with Lachenmann's and Nicolai's highly controlled and controlling methods to engineer what exceeds control and understanding. Kjartansson instead encounters his audience as an unapologetic romantic, fully aware that art may produce intended and unintended consequences for which artists themselves should not be held accountable. His is the work of a melancholic, a cynic, a self-proclaimed nihilist. It is saturated with both sadness and irony; at times borders the silly and kitschy; makes no pretense to mask is imperfection, its improvisational nature, what is haphazard, perhaps even somewhat lazy, sloppy, and indecisive, about it. Kjartansson's performances engage extreme durations, tests everyone's patience, but they rarely come across as representing the rigorous methodological conduct of modern-day scientists or of artists eager to bring art, science, and technology into conversation. Halfway through *Death Is Elsewhere*'s seventy-seven minutes, both Gyða and Kristin Anna Valtýsdóttir suspend their circular stroll and song for a minute or so, urged by their bodies to take a pee break amid the grass. More "rigorous" artists no doubt would have asked their performers to restart the whole performance from scratch to avoid such a pause in the work's final rhythm. In Kjartansson's work, however, such moments themselves become unexpected markers of time, markers of the unpredictable vagaries of matter. They express legitimate claims human affairs and worldly matter in all their inherent messiness make on the putative rigor of aesthetic form, logic, and process. Method, in Kjartansson's work, is detour, just as detour is its central method.

Kjartansson's rhetoric of romantic melancholia differs greatly from the air of technophilia, the aesthetic transposition of scientific modes of inquiry, that speaks to us from *Music for Solo Performer*, *DELPHI Rationale*, and *tele*. His evocation of the profound, the existential, and the metaphysical displays ambitions modern theorists of science such as Max Weber deemed inadmissible to the specialized branches of scientific inquiry.[13] In its unabashedly hyperbolic performativity, Kjartansson's work erases the ground Weber and others considered essential to live up to a scientist's ethos, namely the ability to distinguish between truth and error, right and wrong, even truthfulness and fake. When everything is fake, nothing is—and truth only exists somewhere in between, as a question of gradation, as shady and shifty, as a matter of perspective, interpretation, and contingency, as multivalent, torn, broken, or unresolvable. This, it appears, is not really a good breeding ground for rigorous scientific minds.

[13] Max Weber, "Science as Vocation," in *From Max Weber: Essays in Sociology*, eds. H. H. Gerth and C. Wright Mills (Oxford: Oxford University Press, 1946), 129–58.

But then, again, it is difficult not to think of Rokeby Farm in *The Visitors* as a laboratory, as an experimental space whose separate rooms, wires, earphones, and microphones provide the framework for an ambitious research project. The object of this research is not the resonant vibrations of subatomic matter or the psychophysics of human brain waves. Its object instead is the viability of affect in face of experiences of loss and failure; the connective tissues of human-made and mediated sounds; the vibrant bliss of hospitality amid inhospitable times. *The Visitors* may deeply indulge in romantic melancholia and absorbed pathos, yet the piece is no less than Lucier's 1965 performance dedicated to probing unexpected linkages between interior and exterior worlds, between the perception and the making of sounds, between creativity and a certain yielding of individual agency. And similar to Lachenmann's and Nicolai's projects, *The Visitors* too, in all its stress on unpredictable and non-repeatable outcomes, on intended and unintended consequences, examines a world in which traditional boundaries between matter and mind, subject and object, materiality and consciousness are far from clear.

Kjartansson's aesthetic research emphasizes the unsteadiness and unreliability of knowledge. It advises attitudes that, contrary to modern habits, at times want to slow down our drive toward knowledge or embrace forms of knowing that are and remain partial, inconclusive, hesitant, and murky. It explores forms of knowing no longer hooked to the idea that to know is to grasp something in its entirety, let alone to fully control it and make it our tool, true to the lyrics' at first perhaps most puzzling lines: "There are stars exploding / And there is nothing you can do." Knowledge has been understood as power. But sometimes it might be good to know about the limits of this power as well. Not to know what to do next. To know that, as in the case of interstellar events, our knowing operates on different time scales than what we believe to observe in real time yet may have occurred lightyears ago.

Kjartansson's programmatic endorsement of gestures of waiting and non-interference, of non-programmatic-ness, is instructive, to say the least. To advise quietude in sight of cosmic disasters would surely get any government scientist fired when tasked to find explanations and solutions for pressing problems. But similar to Lucier, Lachenmann, and Nicolai, Kjartansson's research into the enigmas of resonance, affect, telepathic perception, and vibrant matter is not meant to offer rocket science. It explores crosscurrents between art, science, and technological media first and foremost in order to assess the location of art and the aesthetic today. In doing so, however, it cannot but make us think about the question whether more sadness, melancholy, and play, and less stress on data, truth, and utility, could provide today's captains of science and technology with resources to address the calamities of the Anthropocene superior to the ones that led us here in the first place.

Nuclear physics gave us atomic bombs and nuclear fission energy, the historical arrival of both now officially marking the birthyear of the Anthropocene in 1945. But nuclear physics, in the wake of Heisenberg and Bohr, also fundamentally revised our understanding of science: from a practice dedicated to the repeatable verification of hypotheses, the gathering of data, and the analysis of facts in the name of dominating or even re-engineering nature, to a quasi-ecological enterprise in which observer and observed inflect each other, statements of probability replace certainty about predictable outcomes, and the relation of knowing and doing is irresolvably enmeshed. Nuclear physics and quantum mechanics are sad sciences in spite of all of their mind-boggling discoveries. They advise modes of knowing that admit to their own inconclusiveness, their ambivalence and uncertainty, the melancholy of never seeing or grasping anything in its so-called totality or essence, the need to approach something over and over again without ever getting the full picture. Just as they reveal the extent to which matter owns perception and consciousness and may resonate with other matter as capriciously as humans are being affected by a wide range of different events, so they stress that we cannot but live within the very interpretation of what we seek to understand, and that the common distinction between deterministic and fuzzy concepts of resonance may be no more but an illusion entertained by nerdy scientists and over-zealous humanists alike. Scientists after Bohr, rather than explain, interpret the world in ways not so different from how art critics interpret paintings, translators interpret sentences from a different language, and pianists interpret and thereby actualize a written score. "That's what scientists do: they look for patterns in data. Looking at patterns: it's a lot more like appreciating art than you might think."[14]

Lucier's alpha wave sensors, Lachenmann's magical sarod, Nicolai's entangled lasers, Kjartansson's earphones, microphones, and wires—all situate sound and installation art as a laboratory designed to investigate possible resonances between matter and mind, complicate or understanding of the work and place of perception, and thereby establish new relays between science, technological media, and the aesthetic. Is all this, then, a contemporary update of early

[14] Timothy Morton, *Being Ecological* (Cambridge, MA: The MIT Press, 2018), xxvii.

nineteenth-century romantic science,[15] eager to offer radical alternatives to scientific practices too specialized and too much mired in instrumental reason to be able to address big questions of life?

Yes and no.

Resonant knowing is melancholic science. It defies the idea of linear paths from knowledge to doing and values gestures of waiting and delay over frenzied activism. It recognizes that the inconclusiveness of interpretation, in all its different meanings, is all we have and in fact share with the material world of objects. But unlike nineteenth-century romantics, much of this does not result from heroic gestures of rejecting science's cold and self-absorbed reason tout court. It instead owes its momentum from the insight that modern science itself knows more about the (un)logic, unsound, and underland of knowledge than it often is willing to admit. Privileging telepathic uncertainty over causal determinism, melancholic science will not conclusively fix the pressing challenges of the Anthropocene. It at the very least, however, will open our eyes and ears to the fact that we—in the domains of science as much as in the domains of technological development and aesthetic experience—somehow need to learn how to undo our fixation on fixing things, our zeal to police the borders of knowing, in order to develop viable models of resonant survival in the future.

[15] Richard Holmes, *The Age of Wonder: How the Romantic Generation Discovered the Beauty and Terror of Science* (New York: Vintage Books, 2010).

9

Unwalled

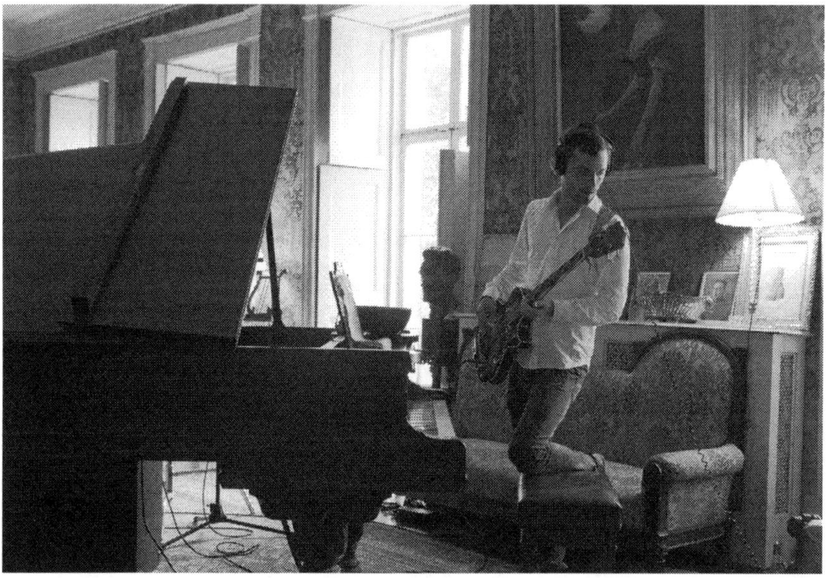

Figure 9.1 Ragnar Kjartansson: *The Visitors*, 2012 (Kjartan Sveinsson). See also Plate 8.

Undoing Loneliness

In the opening pages of *Sonic Warfare*, Steve Goodman calls for a micropolitics of frequency, meant to mobilize affect against contemporary structures of control, oppression, and isolation. Partly banking on vibrations that exceed our bandwidth of hearing, Goodman's politics of sound hopes to undo the functionalization of the audible for the sake of violence and (self-)policing. It promotes new types of entanglements, new modes of hearing and listening, that in Goodman's perspective detach us from the ever-increasing militarization of perception and prioritize openness toward unintended consequences over dominant tactics of coercion and self-management: "The production of vibrational environments that facilitate the transduction of the tensions of urban existence, transforming deeply engrained ambiences of fear or dread into other collective dispositions, serve as a model of collectivity that revolves around affective tonality, and precedes ideology."[1]

Goodman's intervention rests on the understanding of sound and music as sensory spaces dynamically engaged in the making and remaking of social relationships. Audible (or inaudible) vibrations, according to Goodman, collapse binaries that situate us as either performers or listeners, producers or consumers. Sound defines a complex forcefield of echoes and resonances that exceed traditional notions of agency. It provides ecologies of listening that tune and retune how different bodies affect each other, and we judge such ecologies, not by what they may mean or represent, but by the extent to which they enable us to explore, affirm, and celebrate engagements that displace the fearsome orders of the day. Music's meaning is not in what we can say about it, but in how it transforms the bodies from which such saying emanates. Goodman's vibrational environments model alternatives to current experiences of coercion precisely because they, rather paradoxically, rest on the pull of ambient immersion and thereby downplay our ability to choose our surrounds like items on a restaurant menu. No micropolitics of sounds, in his view, can do without questioning the reach of what we have come to hail as the domain of human free will, of selectivity and choice.

With its stress on ceaseless self-management and strategic self-optimization, neoliberal societies of control have created ecologies of being whose privileging of mandatory connectivity is continually haunted by fears about what Hannah

[1] Goodman, *Sonic Warfare*, xx.

Arendt theorized as the modern disaggregation of belonging and community, modernity's systematic production of loneliness. In Arendt's philosophy, states of isolation and solitariness provided fertile grounds for creativity and thought, for our confidence in the vibrancy of the world and our thinking. "In isolation, man remains in contact with the world as the human artifice; only when the most elementary form of human creativity, which is the capacity to add something of one's own to the common world, is destroyed, isolation becomes altogether unbearable."[2] It then slips into loneliness, a state in which we no longer succeed in conforming our identity through the trustworthy company of others, lose trust in the companionship of our thoughts, forfeit our ability to experience self and world, and thus stumble into territories eagerly exploited by terror and totalitarianism. Tyrannies erase the public realms of life by sequestering men's political capacities in architectures of loneliness; they do away with the solitary as a foothold to step into the common and negotiate the infrastructures of life; they integrate the isolated as isolated into spectacles of togetherness only to erect ever-sturdier walls around experience, thought, and communication. Loneliness is totalitarianism's ground zero. It depletes whatever resource is necessary to experience and approach our lives as resonant nodes within dynamic ecologies of relationships, intended effects, and unintended consequences.

There is little common ground between Goodman's sonic re-engineering of the commons and Arendt's stress on solitariness as a precondition for democratic publicness. The latter values the mind's vibrant life—the power of thought and speech—as the portal to non-violent interactions, while the former wants to mobilize affect against the violence we have come to do to ourselves. While the power of Goodman's vibrational environments might transduce fear and thus unburden perception from its functionalization, it does little to prepare the ground for operative public spheres in which humans actualize their potential as rational beings through thoughtful arguments and debates. Arendt, on the other hand, defends the solitary against pervasive loneliness in an effort to ward off the instrumentalization of fear in totalitarian politics, but she has little to say about the affectual dynamics that drive solitary individuals beyond their isolation and energize new attachments to transformative politics. The work of both Goodman and Arendt questions the borders that strategies of coercion erect in order to drain communication, entanglement, solidarity, and critique.

2 Hannah Arendt, *The Origins of Totalitarianism* (1951; New York: Hartcourt Brace and Company, 1976), 475.

Yet they fundamentally disagree about what it takes to remake contemporary ecologies of dread and fear and create more hospitable environments for non-hierarchical interactions.

Or so it seems.

Ecologies know neither of windows nor of outsides. We either find ourselves within them and how they, in however reciprocal terms, structure processes of attention, perception, and communication. Or not. We can tweak ecologies and rearticulate the relational networks from within, but—in contrast to the famous Flammarion engraving—we can't just pass our heads, shoulders, and arms through the limits of one in order to tap halfway into the atmosphere of another. What defines ecologies as ecological is the fact that, in a strict sense, they defy the idea of neutral borders between insides and outsides. They may include different, perhaps even competing areas of relationality, but to think and approach the world ecological is to think of internal boundaries as being in possible flux and external ones as limits to our very ability to sense, speak, think, resound, and communicate.

Goodman's neoliberal age of urban self-management produces environments of fearful isolation as alarmingly as Arendt's age of totalitarianism. Though they would disagree about what kind of micropolitics could challenge the respective politics of containment and isolation, they certainly concur in their assessment of central symptoms of coercion: tears in the fabrics of society so deep in nature that we no longer are able to perceive the interdependence of things and beings, let alone engage or be engaged by the actions of others, without fear, defensiveness, or strategic objectives. Violence, in their respective accounts, not only disrupts unscripted flows of interactions, words, echoes, and reverberations, it not only contains chance, uncertainty, and contingency. As importantly, it obliterates whatever it takes to approach the strange as strange, to be able to coexist with what exceeds our frames of knowledge and understanding, to host within the spheres of our own existence something that isn't us, that isn't of our making, or that may impact us in unforeseeable ways. Violence, in their respective views, defines us and our world as inhospitable. It shuts doors and borders in order to police unpredictable connections and redefine the solitary as lonely. It fuels fantasies of unrestrained sovereignty and demands strategic forms of subjectivity, either of which diminish the principle possibility of seeing and inhabiting the world ecologically.

It is easy to mock Rokeby Farm as a hippie outpost in which trust-fund kids indulge in hedonistic pleasures of shedding fear, urban violence, and political strife, that is, use their privilege to pronounce the advent of a world without. Been there, done that. It is more challenging, however, to argue that Ragnar Kjartansson's *The Visitors* models a world that, on the one hand, demilitarizes perception and mobilizes affect against coercive forms of self-management and that, on the other, embraces the solitary as a critical ground to enable vibrant forms of togetherness. Neither Goodman nor Arendt would, I suspect, think of *The Visitors* as a political work of art. But while I use the word with considerable caution, I want to conclude this book by thinking about what is political about the art of *The Visitors* indeed, not in terms of its overt message or content, but its formal shapes and energies. Kjartansson's *The Visitors* manages to drill gaping holes right into the diamond hearts of our contemporary moment. *The Visitors*' probing of resonant entanglements across distance, its passions, repetitions, and detours, its slowing down of chronological time and haste, its efforts to activate pockets of enchantment and develop scenarios of open-ended attunement, its praise of the power of listening: all this, I suggest, touches viewers and listeners because it has the capacity to transduce neoliberal fear and loneliness, and because it elaborates, in the form of an aesthetic promise, a model of living hospitable amid inhospitable times, of unmaking the very violence that in our contemporary moment informs strategic forms of subjectivity and neo-totalitarian politics of walling.

The borders today's populist nationalisms install to shut out unwanted migrants and strangers mirror the walls tweeting self-optimizers erect around themselves in order to ensure efficient self-realization. In lockstep with other recent aesthetic projects, *The Visitors* is at its most political when it, by exploring the space of art as an ecology of vibrant resonances and unpredictable attachments, brings into play powerful alternatives to contemporary tactics of enclosure. In this, it positions the aesthetic as a medium, an echo chamber, to pattern the affective resources for a future politics of coexistence—a politics which opens our minds and hearts to what is not us and to what may unbuild the many walls we have come to set up to fortify the mansions of our living.

Honing Hospitality

In much of Scottish sound artist Susan Philipsz's work, the echoes of voices and musical sounds transform given spaces, be they indoor or outdoor, into spectral affairs. In *The Lost Reflection* (Figure 9.2), her contribution to the

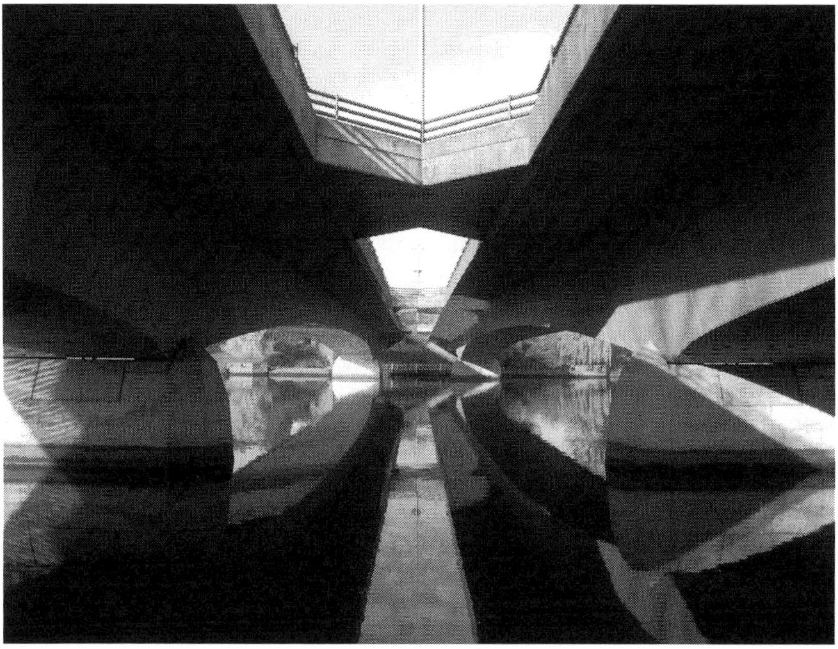

Figure 9.2 Susan Philipsz: *The Lost Reflection*, 2007.

2007 Skulpturprojekte Münster, Philipsz recorded her own voice singing different parts of a duet from Jacques Offenbach's *The Tales of Hoffmann*, allowing them to echo from two different speakers installed on opposite sides of Münster's Lake Aa. Based on the story of Giuletta whose alluring spell caused men to lose their own reflection, Philipsz's resonant score in Münster progressively dissolved what grounds voices in bodies as much as it mirrored a German provincial town uncannily in the distant image of Venice's Grand Canal, the site of the original tale. Philipsz's eight-channel *Study for Strings* deconstructed and reconstructed a score by Pavel Haas, composed in 1943 while being imprisoned in the Nazi concentration camp of Theresienstadt, in order to allow traumatic pasts resonate with forgetful presents. As installed across the tracks of Kassel's historic train station during Documenta 13 in 2012, the echoing sounds of *Study for Strings* sculpted a space able to evoke the presence of an absence as much as they deliberately unmoored the listeners' ability to locate themselves reliably in space and time.

Philipsz's echoes typically involve intricate recording technologies, often splitting and layering Philipsz's own voice across multiple tracks, and they require calculated installation setups in order to achieve their desired effects. The call-and-response patterns of many of her most striking works engage the acousmatic—sounds without visible sources—and in doing so they give voice to what exceeds our immediate grasp. Delicate and fragile, Philipsz's echoes claim the extraterritoriality of beauty in a world haunted by conflict, violence, and trauma. While in some of her work Philipsz's own voice presents itself as the primary medium, her sonic sculptures resonate with the sounds from distant times and other places, so much so that we often seem to leave any spatial compass behind and transcend into crystalline spheres of brittle, but precisely therefore: mesmerizing, wonder. Because death pervades the here and now, Philipsz's echoes break through the walls of the existing world in search of resonant environments elsewhere.

Kjartansson's work is of a different order. It advances a different aesthetic of world-making, of unwalling the fortifications of the present, its specificity becoming the clearest when seen against the backdrop of Philipsz's forays into transcendent beauty. Consider, for a moment, how Heike Munder has described the "unbelievable setting" and "positive heterophony" of *The Visitors* while comparing the installation to some of Kjartansson's earlier works:

> Each part takes place in a different scenery and centered around a different resident of the house. Lost in thought, the figures make music in various rooms, sitting on a sofa or in the bathtub; others have taken their instruments outdoors into the fields. Once again, the individual parts do not coalesce into a whole until they arrive in the exhibition room. A cinematographic tableau emerges that is defined by a profoundly melancholy atmosphere. . . . Gunnarsdóttir's poem, with its subtle sexual charge, addresses the ambivalent roles of femininity; Kjartansson set it to folk music and recorded the piece as a single take at dawn one August morning. Once again, we may recognize the godfathers Stockhausen and Satie in the formal arrangement and the repetitive structures. Everyday and ambient sounds play an important role in the conception: the noise of the tractor and the barking dogs complete the "picture" of the natural setting. To keep the piece with its different protagonists from falling apart, a canon is played every twenty minutes that brings the seven [*sic*] units together for a fleeting moment.[3]

Munder's reference to the ambient qualities of *The Visitors*, its effort to approach and recast music—like Erik Satie—as furniture, its use of iterative elements to

[3] Heike Munder, "Ragnar Kjartansson and the Principle of Nostalgic-Romantic Spatial Music," in *Ragnar Kjartansson: To Music / An die Musik* (Zurich: Migros Museum für die Kunst, 2015), 10.

navigate the heterophony of eight musicians being situated in separated quarters of a house, is illuminating. What deserves further discussion, however, is her stress on the performance's labor to keep things together and achieve some sense of integrated wholeness, as if the piece's primary purpose was to establish and protect borders containing the threat of centrifugal disintegration. While *The Visitors* is indeed about the work of establishing attentive listening as a medium to traverse walls and spatial separations, the concept of space undergirding the installation defies the idea that anything could ever achieve complete closure and integration. Rather than to present space—the space of the performance, the space of the work of art and its reception—as a container whose borders and walls cannot but be haunted by the fear of dissolution, *The Visitors* thinks of space as a site of ongoing emergence, becoming, and transformation, a relational dynamic in which unforeseen links, loose threads, and surprising transgressions—a fizzling out of certain knots in the fabric—are part and parcel of whatever may count as a whole. The "fear" of things falling apart, of seeing connections and relations undone, is the flipside of the pleasures of producing space as a fleeting coming together of diverse voices and improvisational sounds in the present. One requires the other. Each energizes the other. Both support a notion of space, not as a fixed or fixable container of human action, but in Doreen Massey's words, as a "dynamic simultaneity, constantly disconnected by new arrivals, constantly waiting to be determined (and therefore always underdetermined) by the construction of new relations. It is always being made and always, therefore, in a sense, unfinished (except that 'finishing' is not on the agenda)."[4] Whatever vibes and vibrations bring people together will also separate them again; whatever causes us to build walls, borders, and fences to isolate different spaces from each other will generate relational logics that cannot but contest their fixity. No matter how much some may desire integrated wholeness, closure, and unity, nothing—neither in art nor in politics—ever coalesces in such a way that wholes would amount to more than their parts and that borders and walls could really contain for good the movement, becoming, and unfixity of matter. The desire for impenetrable walls and wholeness internally reproduces, in the form of fear, anxiety, and loneliness, what it seeks to shut out from its world. Holes pierce imagined wholes left and right.

You may want to argue, as many have done before, that gestures of hospitality need stable homes and spaces, secure walls and clear separations, a

[4] Massey, *For Space*, 107.

sense of integrated wholeness larger than its parts, in order to grant strangers unconditional welcome. Walls shut out the strange and other, but they also provide the condition for the possibility of refuge, of sheltering itinerant subjects from what troubles, haunts, and threatens them elsewhere. Safe spaces and havens, sanctuaries and asylums—all require some kind of fortification to function as such. Hosts can become hosts only, you may want to conclude, when they control the threshold of different spatial orders. No guest and visitor can enjoy the privilege of meaningful respite without the existence of literal or metaphorical walls segmenting space and arresting the flow of possible movements.

Kjartansson's *The Visitors* urges us to rethink this argument. Whereas Philipsz's sound art seeks to transcend the confines of given places and construct alternate worlds of sonic beauty and uncontained resonance, Kjartansson's work aspires to absorb and be absorbed by—to disappear into—its very spatial setting. Its echoes simultaneously need and perforate the wall that separates our eight musicians. Its beauty lies in building a world within the existing one, yet at the same time this beauty reveals the malleability of walls and spatial structures, the fact that walls are never more than punctuation marks within relational networks of our own making. Just as Arendt cannot think of a functioning public sphere without spaces of and for the solitary, so *The Visitors* envisions a coexistence of human agents across space in which the separate and the common resonate with and cannot do without each other. The work's title is profoundly instructive in this regard and directly linked to how the work considers space as a dynamic simultaneity of relations and walls as media—as interfaces rather than containers—of interaction. For, as much as Kjartansson and his friends enjoy as visitors the hospitality of Rokeby Farm, each of the eight musicians also serves as a resonant host for the distant sounds of the others while the performance of all transforms gallery spaces into sites welcoming audiences to engage in unpredictable acts of listening and viewing. Hosts are hospitable hosts, we learn from Kjartansson, not because they own fortified houses and command impermeable walls, but because they know all too well that we all are guests and visitors in some place, at some time, according to some perspective. The difference between hosts and visitors is neither ontological nor certain. It is performative, a function within a larger network of movements, positionings, impacts, and effects. It is, if you wish, Heisenbergian. You are never just one or the other. When seen from different observational standpoints, you might be both at once. Hosts are good hosts, according to *The Visitors*' choreography of echoing calls and responses, not only because they know that walls are nothing but screens and projections that by themselves do not afford the closure and stability we typically assign to them, but because they for this reason understand how to coexist with what is visiting and strange in themselves.

Paraphrasing a well-meaning saying about foreigners in xenophobic times, you may say that if everyone's a visitor somewhere nobody should be considered as one at all. The formal organization and moving beauty of Kjartansson's *The Visitors* tells us something else, something more radical, though. If being visitors and migrants is part of the human condition, then nothing—according to the aesthetic intervention of *The Visitors*—appears more important than to hone our hospitality amid inhospitable spaces. Which means, among others: to reimagine walls as relational media rather than as obstacles to possible attachments; to find, explore, and develop resonant environments and echo chambers that disentangle traditional binaries between hosts and visitors; and to learn, however melancholically, how to flee and fall into our dark times rather than propagate the (impossible) construction of alternate worlds elsewhere. The work of two other contemporary sound and installation artist to which we now briefly turn is helpful to sketch this out further and flesh out the full political import of Kjartansson's aesthetic of resonant hospitality.

Beyond Hunkering and Huddling

"Walls built around political entities," writes Wendy Brown, "cannot block out without shutting in, cannot secure without making securitization a way of life, cannot define an external 'they' without producing a reactionary 'we,' even as they also undermine the basis of that distinction. Psychically, socially, and politically, walls inevitably convert a protected way of life into hunkering and huddling."[5] As much as walling in recent years has served to block the influx of unwanted bodies, subjectivities, and workers, it first and foremost functions theatrically. It performatively seeks to express the very power and centralized authority global flows of moneys, work, technologies, bodies, values, and ideas tend to undermine. The walls, fences, and enclosures of our present display the work of rampant desires to reinstate forms of geopolitical sovereignty that, in the post-Westphalian order of

[5] Wendy Brown, *Walled States, Waning Sovereignty*, with a new preface (2010; New York: Zone Books, 2017), 54.

modern Europe and beyond, sought to inject theological notions of divine power and control into political realities. Donald Trump's US-Mexican wall, Israel's security fence enclosing Palestine territories, Viktor Orbán's Hungarian border fence—all offer material structures to screen out unwanted realities, as much as they provide surfaces to screen visions of sovereignty. Materially far less successful than their advocates want them to be, the primary purpose of the enclosures of our time is to entertain fantasies about distinguishing friends and enemies, us and them, and precisely thereby reinvoke forms of political authority for national or even supranational entities that have been weakened by these entities' very pursuit of unhampered trade, investment, securitization, and debt.

Lawrence Abu Hamdan's one-channel video installation *Walled Unwalled* (2018) (Figure 9.3) offers a compelling example of how contemporary art can

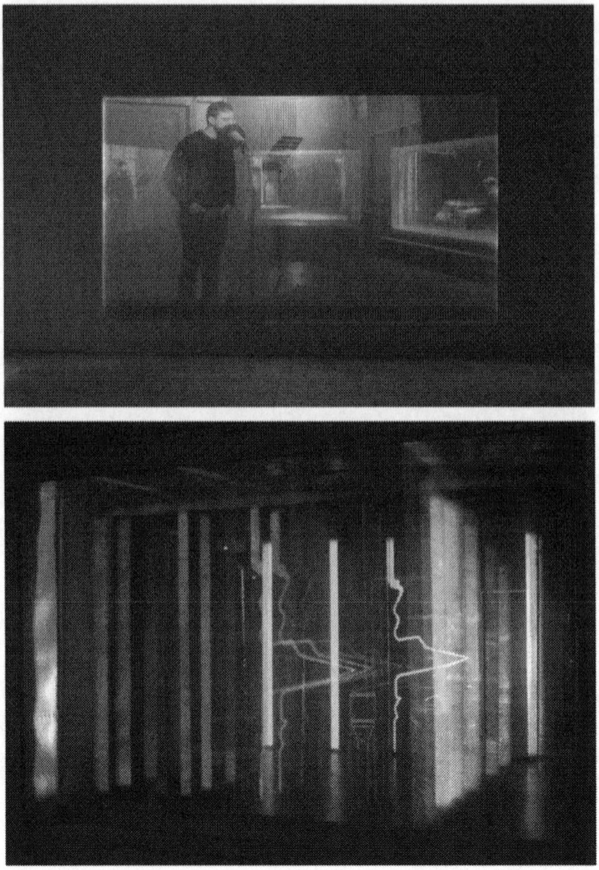

Figure 9.3 Lawrence Abu Hamdan: *Walled Unwalled*, 2018.

investigate the psychopolitics of our age of walling, but also unsettle the very hunkering and huddling left in the wake of projective securitization. *Walled Unwalled* is dedicated to a series of judicial cases in which physical walls confounded the legal privileging of visual evidence and assigned the audible a key role in reconstructing the past. Recorded in a radio station in Berlin that had once played a critical role in legitimizing the Berlin Wall, Abu Hamdan's performative intervention recalls, among other, the murder case against the South African Olympian Oscar Pistorius and the experiences of former inmates in the Syrian torture prison Saydnaya. In the first case, central defense strategies rested on what Pistorius believed to be hearing through the wall of his bathroom when he shot his girlfriend Reeva Steenkamp in 2013. Abu Hamdan's forensic reconstruction leaves little doubt that Pistorius's claims were incompatible with the possible propagation of sound waves and reverberations. A different segment, on the other hand, asks Syrian torture victims to retell their harrowing imprisonment and map their prison's dark structures through auditory memories. Various former prisoners describe tormenting tremors of the entire building, explained by Abu Hamdan as effects of torture practices whose sounds caused a buildup of standing air columns in the prison's observation towers and resonant vibrations in each cell's walls. While the camera pans back and forth between different individual recording studios, Abu Hamdan—in a third case study—cites the words of an Oregon cannabis grower whose product was said to have extraordinary effects. "I don't know what's behind the wall. I know no longer there is a wall. I know no longer this wall is a wall. I know no longer what a wall is. I no longer know that in my room there are walls, but if there weren't any walls there would be no room." The farmer's memories of intoxication uncannily anticipate what has become the order of the day in our twenty-first century of populist walling and neoliberal loneliness: an age eager to erect border fences at all cost, even though their physical function lacks effectiveness and their symbolic role all too clearly demonstrates the work of complex logics of displacement and projective mechanisms. In Abu Hamdan's own words: "Today we are all wall and no wall at all."

Abu Hamdan's walls separate and isolate. They sequester bodies, physical movements, perceptions, and thoughts. Though not entirely impermeable, they are agents of violence, play a critical role in the policing and self-policing of subjectivities, in controlling affects and securitizing communication. And yet, even though resonant sounds may pass the walls of the present and, from an epistemological standpoint, complement the limitation of the visual, it would be foolish to think that acoustic passages per se can undo the violence and loneliness administered by contemporary walling. On the contrary. For walls, as Abu Hamdan's drug farmer knows all too well, represent physical as much as psychological entities; they live in us and mess with our affects as controllingly as they structure architectural, social, and political spaces. And, as Abu Hamdan's former inmates of the Saydnaya complex will add, resonant vibrations have the potential to amplify rather than dissolve terror and coercion. Tearing down or permeating physical walls with sounds per se will therefore do very little to set subjectivities free and enable new types of non-hierarchical entanglements. It takes much more than trumpets to overcome the huddling and hunkering—the walls that split apart perceptions, emotions, and minds—in our age of securitization, surveillance, and self-management. And yet, resonant sounds may take the lead for us to rethink and recalibrate notions of identity, subjectivity, and agency whose uncontrolled application empowers walling advocates, in the name of increased control and safety, to erect ever-sturdier fortifications, to develop ever-more effective structures of loneliness and coercion.

As a potent example of this rethinking, consider the pathbreaking work of composer, experimental musician, and sound and installation artist Guillermo Galindo, including his most iconic *Ángel exterminador*, first exhibited in 2015 (Figure 9.4). *Ángel exterminador* consists of a twisted iron section of the US-Mexican border wall, found and picked up in Jacumba, California, and then carefully attached to a wooden scaffold with the aim to create both a sculptural object and a musical instrument. The sculpture's piece of metal mimics the wings of an angel or the corpse of a dead man. It is designed to be played like a gong, and its trembling is meant to evoke stories of loss and hardship often rendered invisible. *Ángel exterminador* emerged from a collaboration between Galindo and photographer Richard Misrach, a multimedia project entitled *Border Cantos*. Misrach documented numerous sections of the US-Mexican border in photographs as stunningly engaging as shockingly surreal. Galindo amassed objects left and lost by passing migrants to handcraft unique musical instruments, and composed, performed, and recorded various sound works whose acoustical shapes and energies infused his sad collection of objects with new life—a second life singing stories of agony and despair, but also of perseverance, obstinacy, and hope.

"When designing instruments," Galindo describes his practice,

> my goal is not to obtain the perfect or most beautiful sound, but to allow the materials to sing in their own voices. There are many reasons why I refuse to

> consider my pieces recycled art objects. The instruments for the Cantos project are meant to enable the invisible victims of immigration to speak through their personal belongings. Using their own narrative, these instruments tells us imaginary stories about places and people that may or may not still be alive. Other instruments for this project came from the apparatus of division itself. These objects of aggression were also given a new life and an opportunity to speak in their own terms.[6]

Galindo's *Ángel exterminador* is not a melancholic angel of history, one who—like Walter Benjamin's—has become mute in face of the disasters of time. His mission is not simply to hold on to the debris of history so as to preserve it for a future moment of redemption and resurrection. Instead, *Ángel exterminador* aspires to beat violent history at its own game. Galindo's sculptures and instruments enable seemingly ordinary things to resound in unexpected ways and thereby not only to dispute the silencing of precarious voices in the present, but to profile the extent to which our contemporary politics of

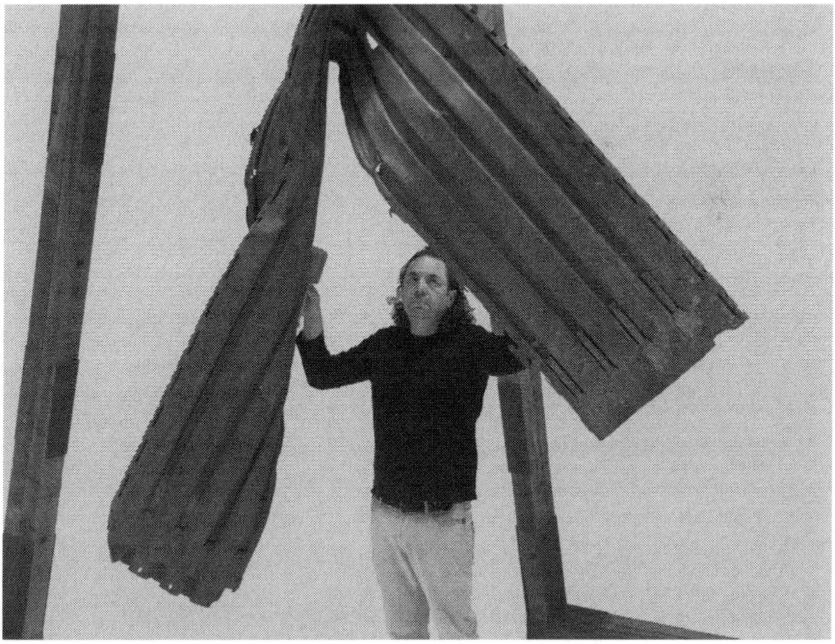

Figure 9.4 Guillermo Galindo: *Ángel exterminador (Exterminating Angel)*, 2015.

[6] Guillermo Galindo, "Sonic Borders," in *Border Cantos*, by Richard Misrach and Guillermo Galindo (New York: Aperture, 2016), 193.

exception have rendered ordinariness elusive. Galindo's *Ángel exterminador* is the angel of resilient counter-histories. Despair and hope here fold into one single dynamic.

The sound of music, for Galindo, far exceeds what can resound in a concert hall, emanate from a radio's loudspeaker, or be represented on a traditional score sheet.[7] To compose music, according to Galindo, is to echo and rechannel the real's storehouse of possible und fundamentally unpredictable sonic events. It is to lend sound to the seemingly silent, but in no way mute passive objects and material makeup of the world; it is to collect inaudible things and gather physical inscriptions so as to translate their latent voice into something that may touch upon the human ear and modulate our affect. To play music means to play the world as much as it is to invite the world to play on us; it means to probe possible resonances between the human and the nonhuman, not to establish harmonious consonance, but to allow sound to make unexpected connections and thereby, in spite and in face of the unbearable burdens and violence of the present, restore some sense of contingency, attachment, and ordinariness.

Though many of Galindo's musical instruments and performances require both the hands of the artist and the operations of mechanical recording devices, the sounds Galindo teases out of his collected objects relate to matter as if it had vitality, a voice, of its own, as if to compose meant nothing but to translate the will, vibration, and agential power of things into a register—a frequency—audible to the human ear. As a ragpicker of what the twenty-first century's precariat of migrants and refugees are forced to leave behind, Galindo's work taps into resonant sounds to push against a human world that—as explored by Abu Hamdan—eagerly divides space into separate realms of friends and foes, humans and nonhumans, subjects and objects. He transforms our precarious present into an aesthetic echo chamber powerful enough to unbuild the very

[7] Here's how Galindo himself describes his initial impulse to collect and recompose the material traces and trails of American walling policies: "The first time I saw Richard Misrach's photograph of a Border Patrol tire drag, with the lines on either side of a sandy desert road, I thought about diagonal staff lines in motion, pulling the viewer into the picture and/or the listener into a sonic sequence. From the beginning I thought of the process of translating the data in the photographs into scores, in the same way that a computer converts arrays of zeroes and ones into pictures, music, or words. In my scores I also tried not to exclude the emotional elements, the narrative and the archetypal symbolism: people behind bars, a flock of birds, an empty sky, a surveillance tower, gun shells, etc. Translating photographs into scores also allowed me to challenge the traditional Cartesian tradition of reading music from left to right. These photographs inspired me to think of unexpected events coming from all directions all at once or at different times: a polycentric universe where anything can happen at any given time" (Galindo, "Sonic Borders," 195).

walls contemporary calls for securitization erect right in our minds and hearts—walls that make us hunker and huddle in isolation, undermine our ability to both listen and speak up in public, and prevent us from resisting ubiquitous mechanisms of coercion that produce the contemporary as a deeply inhospitable space.

Dwelling in Sound, Dance, and Art

The purpose of architecture, writes Norwegian architectural theorist Christian Norberg-Schulz, is "to make a site become a place, that is, to uncover the meanings potentially present in the given environment."[8] Drawing on Heidegger's concept of dwelling, he adds that any enclosure is defined by its boundary and any boundary is determined by its openings, its windows and doors—thresholds at which something begins its presencing. "In general the boundary, and in particular the wall, makes the spatial structure visible as continuous or discontinuous extension, direction and rhythms."[9] Walls emerge as walls, and buildings as buildings, only to the extent to which their openings define the character of the relationship between interior and exteriors spaces and gather the landscape around the dwelling as landscape.

The walls of Rokeby Farm, and of the gallery spaces that in various international locations have hosted Ragnar Kjartansson's *The Visitors* and its audiences, at once echo and transcend this idea of dwelling and inhabitation for at least two reasons. First, they are necessary to envision the commons as a site for the generative coexistence of the solitary rather than the lonely, yet they neither come in the singular, nor do they exhibit a clear sense of enclosure, boundary, and opening, of what may count as inside and what as outside, what is wall and what is window. Rokeby Farm's walls are membranes simultaneously separating and connecting individual musicians within a larger vibrational environment of transductions, while the gallery's screens and speakers, as they afford roaming audiences to experience and probe ever-different portals of immersion, vastly complicate and in fact unbuild the power of fenestral framing. Second, the walls featured in *The Visitors* might be necessary to anchor perception, and to transform space into place and place as a site endowed with character, ambience,

[8] Christian Norberg-Schulz, *Genius Loci: Towards a Phenomenology of Architecture* (New York: Rizzoli, 1980), 18.

[9] Norberg-Schulz, *Genius Loci*, 13.

and a certain "genius." But the logic of the nine-channel video and its installation leaves little doubt that, in contrast to Heideggerian notions of dwelling, we cannot access the genius of a place in our world of electronic media, mobility, and migration without bringing into play the relational affects and networks of space, the entanglement of things local and global, interior and exterior. Which is just another way of saying that *The Visitors* urges us to rethink notions of dwelling, enclosure, and hospitality for a twenty-first century world in which Heidegger's hut in the woods can no longer serve as a viable model of being and even the most strenuous effort of thought will fail to neatly separate concrete from abstract locations, lived place from relational space. If, today, we—as Abu Hamdan claims—are all wall and no wall at all, then the acoustic modeling of resonant ecologies in Kjartansson's or Galindo's work is much more appropriate to approach the question of how we can coexist with others than dominant frameworks informed by the visual dialectics of wall and window, enclosure and opening.

Dwelling, in Heidegger's world, is meant to gather the world, with the help of structures of enclosure, into meaningful horizons around the pivoting figure of the human. It offered an existential foothold within the concrete fabrics of the everyday catering to the human need for identification and orientation, for belonging: to build walls and dwell within them was to make friends with a particular environment. Kjartansson's *The Visitors*, as much as Galindo's *Border Cantos*, recast dwelling from the standpoint of sound and the auditory. Both downplay the concept's historical stress on the centrality of meaning-making and goal-oriented human subjects, and instead explore the possibilities of attuning to environments that exceed our understanding, control, and intentionality. Echoes need walls to bounce, yet in bouncing they fundamentally reshape our maps of space, our sense for the limits and textures of the sensible. Just as we find ourselves always already within landscapes of sound and unsound, of mediated vibrations and vibrant media, so our modes of dwelling need to account for the fact that we always are hosts and guests at once, belong and do not belong; that we are just one moving node and membrane within a larger nexus of ever-shifting relations of vibrant matter. The point of architecture is not to gather the world around us in order to confirm and stabilize preexisting notions of interiority and exteriority, subject and object, the commons and the solitary. It is not about securing enclosures of individual belonging, but about offering media of resonant interactions between different agents. It is about developing sites at which the strange, other, and incommensurable can coexist with and affect each other without retreating into mindless fear or building up ever-more defensive shields.

The question, then, is inevitable: isn't this acoustic retuning of dwelling as a place of resonant hospitality helplessly utopian and romantic? Isn't Kjartansson's, and for that matter: Galindo's, desire to explore ecologies in which different bodies, agents, and temporalities, the solitary and the common, the distant and the proximate, can fearlessly coexist and fall into each other a mere aesthetic pipedream and hence deeply out of touch with political and social realities?

In the late 1960s Theodor W. Adorno ruminated that nothing about contemporary art is self-evident anymore, including its very right and legitimation to exist.[10] For art, to define itself as art, it needed to ceaselessly question its own location in society and recognize its critical rejection to blur into social life as a social fact itself. Yet rather than—as Adorno on some level feared—see artworks disappear in the wake of their own refusal to communicate, the last few decades have largely come to consider art as a mere replica or direct feeding ground of the social. Many are deeply suspicious that art, by taking on incalculable risks and unintended consequences, could violate the expectations for thought and action beyond its purview. Others are profoundly impatient about the fact that art may only be able to become and enjoy the privilege of art by not being not-art. A third set of critics, finally, may proclaim a potent return of aesthetic experience, yet in arguing that the aesthetic—the distribution of the sensible—subtends all things political seamlessly dissolve art's otherness into the everyday, the social, through the backdoor. In no longer wanting to grant aesthetic practices relative autonomy from the exigencies of the real, what we—forever suspicious—have largely lost in the shuffle is our ability to approach art as an environment in which objects or figures of movement are welcome to exert unexpected pressures on body and mind causing viewers to surrender to forces not of their choice. In expecting the aesthetic itself to live up to the imperatives of ceaseless self-management and strategic interactivity, what we have lost out of sight is the fact that art is at its best whenever it defies the protocols and (un) logics of the everyday, claims some room for its own, and invites us to probe unlikely cohabitations of the strange and incommensurable; whenever it lures us into unexpected encounters for which no prior concept exists and urges us, perhaps even coerces us, to lower our defensive shields; whenever it makes us

[10] Theodor W. Adorno, *Aesthetic Theory*, trans. Robert Hullot-Kentor (London: Continuum, 2002): "It is self-evident that nothing concerning art is self-evident anymore, not its inner life, not its relation to the world, not even its right to exist" (1).

resonate with something that isn't us or of our making and precisely thereby helps us recognize that our everyday—its mania of walling, its hunkering and huddling—essentially rests on the very barring of resonance and play, of hospitality and fearlessness; whenever it, in short, invites us to attune our senses and minds to experiences of startling beauty, speechless surprise and wonder, or daunting disorientation.

Attunement, writes Timothy Morton,

> is a dance between completely becoming a thing, the absolute camouflage of pure dissolution (one kind of death) and perpetually warding off that thing (another kind of death), the mechanical repetition that establishes walls, such as cell walls. Between I am that and me me me: in other words, between being reducible to other stuff (I'm just a pile of atoms or mechanical components) and being totally different from other stuff (I'm a person, and only some beings get to be people).[11]

Art today cannot but register and take on the coercive hunkering and aggressive huddling that represent the principal signature of our times, yet it can only succeed in doing this if it learns how to dance, play, and resonate with what exists beyond intention, sovereignty, and purpose. When it gives in to ubiquitous pressures of either doing more or doing less than it really can or is prepared to do, when it simply aspires to offer either political action or individual self-help, pedagogy or atmospheric gratification, social work or regenerative wellness—then contemporary art is bound to fail at hosting the dance between different forms of death. The romantic in art today approaches matter as vibrant and as more than merely a gathering of numb things, as much as it defies the excesses of neoliberal me-culture, the hypertrophies of strategic subjectivity and protective isolationism. It may well be that art today has no choice other than to endorse some version of the romantic lest it wants to lose its reason and right to exist indeed.

Echo's Promise

And this finally returns us to the question that set this whole book into motion: What is it about Ragnar Kjartansson's *The Visitors* that moves audiences such that they want to watch its 64-minute loop over and over again? What kind of

[11] Morton, *Being Ecological*, 111.

tears tear through the order of the everyday when we find ourselves entangled with the sounds emanating from the installation's nine speakers, attached to its individual screens, for minutes and perhaps even hours on end? Each chapter of this book has added another piece to the puzzle, traversing a variety of different echo chambers to resound possible responses. All together propose to find the answer in nothing more, and nothing less, than the profoundly romantic art and aesthetics of "falling," as praised in the lyrics of Ásdís Sif Gunnarsdóttir, and as practiced in the effort of Kjartansson and his friends, through attentive listening and echoing music making, to establish resonant relationships across Rokeby's walls.

Few languages, at least Indo-European ones, tether positive meanings to the act of falling. Just as their vocabularies prefer what is above ground to what resides below our feet, so they typically valorize the activity of uprightness, deliberate movement, and willful directedness over the cessation of control—the presumed passivity—of falling. Gunnarsdóttir's "Once again I fall into my feminine ways" speaks a different language. Her engendering of the figure of falling might raise eyebrows with some, galvanize others. I here simply want to see it as a tribute to what literal or metaphorical experiences of falling within the space of art can do, namely, to unbuild the hardened walls of strategic reason and combative self-management, to rebuild and probe the possibility of resonating with our environments amid painful times, of playing and dancing, of living and being ecological.

The idea of "falling" in Gunnarsdóttir's poem—and, by extension, in Kjartansson's performance is not about renouncing agency and action for the sake of essentializing concepts of passive femininity. The English language knows more about this than some of its Indo-European cousins. After all, in English, the notion of "falling in love" describes sudden experiences that wash over us without our doing and control, but it also registers unexpected emotional attachments that animate the subject, in all its vulnerability, in all kinds of new ways. And there is more. Calling it the happiest thought of his life, Albert Einstein wrote in 1907 that, "For an observer falling freely from the roof of a house there exists—at least in his immediate surrounding—no gravitational field."[12] Similar to how acts of falling undo our sense of boundaries and unsettle fixed measures of movement, speed, and identity, the poetic coupling of falling

[12] Abraham Pais, *Subtle Is the Lord: The Science and the Life of Albert Einstein* (Oxford: Oxford University Press, 1982), 178.

and the feminine in Gunnarsdóttir and Kjartansson envisions conditions under which tenacious binaries between activity and passivity, doing and receiving, host and visitor, interior and exterior, the solitary and the communal can lose their bearing. To fall into one's feminine ways, in *The Visitors*, means to recognize the limits and fragility of human meaning-making, world-making, and wall-making. It recalls and reanimates the forgotten promise of a world different from the existing one in which we can let go and become other, echo the strange in us and resound the alien, begin to dance and expose our fragility, embrace chance and invite the contingent, all without fearing harm or anticipating retribution.

Kjartansson's art is unapologetically romantic and melancholic, yet precisely in playing out affects, desires, and reverberations that have no place amid the coercion of the everyday offers a sundial of our precarious times. To fall into the feminine, to fall into one another's sounds and music, is not only to tune into what exceeds our making, our sense of control, but in doing so also to chart the very logic that keeps us from doing so in everyday life. By seemingly wanting to exist entirely for its own sake, art here transduces unfulfilled claims and promises of happiness against the coercive routines and inhospitable shapes of the present. Resonance has no intrinsic meaning. It can liberate the individual from the confines of identity, but it can also subject the subject to destructive forces. It unbuilds walls, lets in the other without conditions or demands, yet it may also obliterate what permitted entry in the first place. *The Visitors* invites the viewer to fall in, play through, and dance with the ambivalences of resonant vibrations within the hospitable space of art. It brushes against the confines of neoliberal loneliness and walling at the level of form without giving up on the solitary as a ground of meaningful community. By—romantically, indeed—inhabiting art and the aesthetic as vibrant ecologies of their own, the installation extends promissory notes for a future in which we no longer need to fear that our entanglements with the world's matter might destroy us.

How is it possible, then, that *The Visitors*, like no other sound and video installation of our contemporary moment, succeeds in touching the lives of its audiences and even moves many of its listeners to unexpected tears? Because its unconditional trust in the romantic in art, its overwhelming beauty, reminds us that we cannot follow Echo's desire to live a life for the sake of love, of vibrant attachment, if we fail to retune our times' inhospitality and, in spite of all odds, melt down our present's diamond hearts. Because its resonant sounds and echoes leave us with no other choice but to redesign our ecologies of being and, by taking the playful in art seriously, to change our lives.

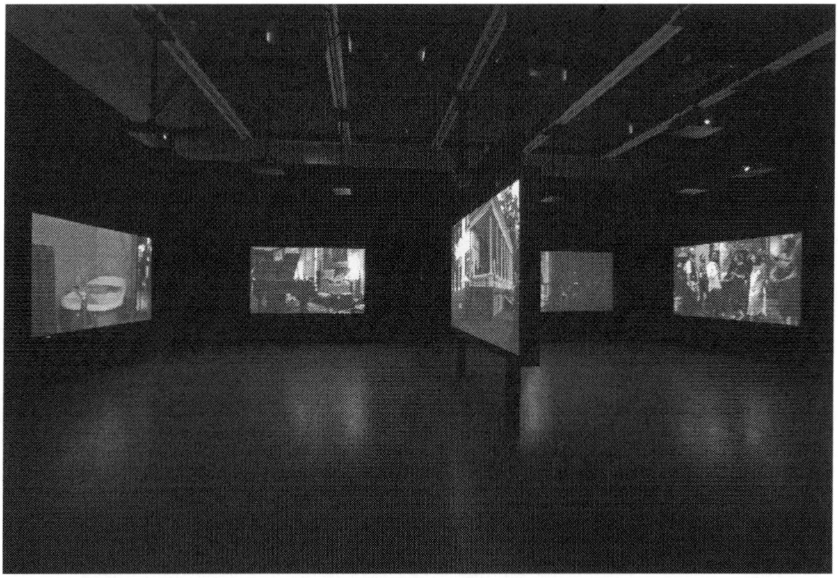

Figure 9.5 Ragnar Kjartansson: *The Visitors*, 2012. Installation view at Luhring Augustine, New York, 2013.

Selected Bibliography

Adorno, Theodor W. *Aesthetic Theory*. Translated by Robert Hullot-Kentor. London: Continuum, 2002.

Arendt, Hannah. *The Origins of Totalitarianism*. 1951. New York: Hartcourt Brace and Company, 1976.

Ball, Hugo. *Flight Out of Time: A Dada Diary*. Translated by Ann Raimes. New York: Viking, 1974.

Balsom, Erika. *Exhibiting Cinema in Contemporary Art*. Amsterdam: Amsterdam University Press, 2013.

Barrett, G. Douglas. *After Sound: Toward a Critical Music*. New York: Bloomsbury, 2016.

Barthes, Roland. *Image—Music—Text*. Translated by Stephen Heath. New York: Hill and Wang, 1977.

Bennett, Jane. *The Enchantment of Modern Life: Attachments, Crossings, and Ethics*. Princeton: Princeton University Press, 2001.

Bennett, Jane. *Vibrant Matter: A Political Ecology of Things*. Durham: Duke University Press, 2010.

Berlant, Lauren. *Cruel Optimism*. Durham: Duke University Press, 2011.

Bernstein, J. M. *Against Voluptuous Bodies: Late Modernism and the Meaning of Painting*. Stanford: Stanford University Press, 2006.

Bianchi, Frederick and V. J. Manzo, Eds. *Environmental Sound Artists: In Their Own Words*. Oxford: Oxford University Press, 2016.

Bishop, Claire. *Installation Art: A Critical History*. London: Tate, 2005.

Bishop, Claire. *Artificial Hells: Participatory Art and the Politics of Spectatorship*. London: Verso, 2012.

Blackbourn, David. *The Conquest of Nature: Water, Landscape, and the Making of Modern Germany*. New York: W.W. Norton, 2006.

Blesser, Barry and Linda Ruth-Salter, Eds. *Spaces Speak, Are You Listening? Experiencing Aural Architecture*. Cambridge, MA: MIT Press, 2006.

Böhme, Gernot. *Atmosphäre*. Frankfurt am Main: Suhrkamp, 1995.

Brown, Wendy. *Walled States, Waning Sovereignty*. With a new preface. New York: Zone Books, 2017.

Bubaris, Nikos. "Sound in Museums—Museums in Sound." *Museum Management and Curatorship*, 29.4 (2014): 391–402.

Byrne, David. *How Music Works*. New York: Three Rivers Press, 2012.

Cage, John. *Silence: Lectures and Writings by John Cage*. Middletown, CT: Wesleyan University Press, 1961.

Cahan, David. *Helmholtz: A Life in Science*. Chicago: University of Chicago Press, 2017.

Cavarero, Adrianna. *For More than One Voice: Toward a Philosophy of Vocal Expression*. Translated by Paul A. Kottmann. Stanford: Stanford University Press, 2005.

Chayka, Kyle. *The Longing for Less: Living with Minimalism*. New York: Bloomsbury, 2020.

Chion, Michel. *Audio-Vision: Sound on Screen*. Edited and translated by Claudia Gorbman. New York: Columbia University Press, 1994.

Chion, Michel. *Sound: An Acoulogical Treatise*. Translated by James A. Steintrager. Durham: Duke University Press, 2016.

Chladni, Ernst Florens Friedrich. *Entdeckungen über die Theorie des Klanges*. Leipzig, 1787.

Cixous, Hélène. "The Laugh of Medusa." Translated by Keith Cohen and Paula Cohen. *Signs*, 1.4 (Summer 1976): 875–93.

Cluett, Seth. "Ephemeral, Immersive, Invasive: Sound as Curatorial Theme, 1966–2013." In *The Multisensory Museum: Cross-Disciplinary Perspectives on Touch, Sound, Smell, Memory, and Space*. Edited by Nina Levent and Alvaro Pascual-Leone. Lanham, MD: Rowman & Littlefield, 2014, 109–18.

Connor, Steven. *Dumbstruck: A Cultural History of Ventriloquism*. Oxford: Oxford University Press, 2000.

Connor, Steven. "Ears Have Walls: On Hearing Art." Lecture given in the series *Bodily Knowledges: Challenging Ocularcentricity*. Tate Modern, February 21, 2003. http://www.stevenconnor.com/earshavewalls/.

Connor, Steven. "The Strains of the Voice." In *Phonorama: Eine Kulturgeschichte der Stimme als Medium*. Edited by Holger Wölfle. Berlin: Matthes and Seitz, 2004, 158–72.

Cox, Christoph. *Sonic Flux: Sound, Art, and Metaphysics*. Chicago: University of Chicago Press, 2018.

Cox, Christoph and Daniel Warner, Eds. *Audio Culture: Readings in Modern Music*. New York: Bloomsbury, 2017.

Crary, Jonathan. *24/7: Late Capitalism and the Ends of Sleep*. London: Verso, 2013.

Deleuze, Gilles and Felix Guattari. *A Thousand Plateus: Capitalism and Schizophrenia*. Translated by Brian Massumi. Minneapolis: University of Minnesota Press, 1987.

Demers, Joanna. *Listening through the Noise: The Aesthetics of Experimental Electronic Music*. Oxford: Oxford University Press, 2010.

Denning, Michael. *Noise Uprising: The Audiopolitics of a World Musical Revolution*. London: Verso, 2015.

Derrida, Jacques and Anne Duforurmantelle. *Of Hospitality*. Translated by Rachel Bowlby. Stanford: Stanford University Press, 2000.

Dolar, Mladen. *A Voice and Nothing More*. Cambridge, MA: MIT Press, 2006.

Duffy, Enda. *The Speed Handbook: Velocity, Pleasure, Modernism*. Durham: Duke University Press, 2009.

Dumbadzse, Alexander. *Bas Jan Adler: Death Is Elsewhere*. Chicago: University of Chicago Press, 2013.

Dyson, Frances. *Sounding New Media: Immersion and Embodiment in the Arts and Culture*. Berkeley: University of California Press, 2009.

Dyson, Frances. *The Tone of Our Times: Sound, Sense, Economy, and Ecology*. Cambridge. MA: MIT Press, 2014.

Egström, Andreas and Åsa Stjerna. "Sound Art or *Klangkunst*? A Reading of the German and English Literature on Sound Art." *Organised Sound*, 14.1 (2009): 11–18.

Eidsheim, Nina Sun. *Sensing Sound: Singing and Listening as Vibrational Practice*. Durham: Duke University Press, 2015.

Einstein, Albert, Boris Podolsky and Nathan Rosen. "Can Quantum-Mechanical Description of Physical Reality Be Considered Complete?" *The Physical Review*, 47.10 (1935): 777–80.

Erlmann, Veit. *Reason and Resonance: A History of Modern Aurality*. Cambridge, MA: MIT Press, 2010.

Erlmann, Veit. "Resonance." In *Keywords in Sound*. Edited by David Novak and Matt Sakakeeny. Durham: Duke University Press, 2015, 175–82.

Farland, David. *Drawing on the Power of Resonance in Writing*. Scotts Valley: CreateSpace, 2013.

Fay, Jennifer. *Inhospitable World: Cinema in the Time of the Anthropocene*. Oxford: Oxford University Press, 2018.

Fink, Robert. *Repeating Ourselves: American Minimal Music as Cultural Practice*. Berkeley: University of California Press, 2005.

Fisher, Philip. *Wonder, the Rainbow, and the Aesthetics of Rare Experiences*. Cambridge, MA: Harvard University Press, 1998.

Goodman, Steve. *Sonic Warfare: Sound, Affect, and the Ecology of Fear*. Cambridge, MA: MIT Press, 2012.

Gruneisen, Peter and Bob Hodas, Eds. *Soundspace: Architecture for Sound and Vision*. Basel: Birkhäuser, 2003.

Gumbrecht, Hans Ulrich. *In Praise of Athletic Beauty*. Cambridge, MA: Belknap Press, 2006.

Hankins, Thomas and Robert Silverman. "The Aeolian Harp and the Romantic Quest of Nature." *Instruments and the Imagination*. Princeton: Princeton University Press, 1995, 86–112.

Heidegger, Martin. *Poetry, Language, Thought*. Translated by Albert Hofstadter. New York: Harper Colophon Books, 1971.

Helmholtz, Hermann von. *On the Sensations of Tone as a Physiological Basis for the Theory of Music*. Translated by Alexander J. Ellis. New York: Dover Publications, 1954.

Holmes, Richard. *The Age of Wonder: How the Romantic Generation Discovered the Beauty and Terror of Science*. New York: Vintage Books, 2010.

Hörisch, Jochen. *Ende der Vorstellung: Die Poesie der Medien*. Frankfurt am Main: Suhrkamp, 1999.

Jackson, Myles W. *Harmonious Triads: Physicists, Musicians, and Instrument Makers in Nineteenth-Century Germany*. Cambridge, MA: MIT Press, 2006.

Jankélévitch, Vladimir. *Music and the Ineffable*. Translated by Carolyn Abbate. Princeton: Princeton University Press, 2003.

Jenny, Hans. *Cymatics: A Study of Wave Phenomena and Vibration*. Eliot, ME: MACROmedia Publishing, 2001.

Jones, Mari Riess. "Only Time Can Tell: On the Topology of Mental Space and Time." *Critical Inquiry*, 7 (1981): 557–76.

Jullien, François. *In Praise of Blandness: Proceeding from Chinese Thought and Aesthetics*. Translated by Paula M. Varsano. New York: Zone Books, 2008.

Kahn, Douglas. "Alvin Lucier: *I Am Sitting in a Room*, Immersed and Propagated." *Oase*, 78 (2009): 24–30.

Kahn, Douglas. *Earth Sound Earth Signal: Energies and Earth Magnitude in the Arts*. Berkeley: University of California Press, 2013.

Kandinsky, Wassily. *Complete Writings on Art*. Edited by Kenneth C. Lindsay and Peter Vergo. New York: Da Capo, 2004.

Kane, Brian. *Sound Unseen: Acousmatic Sound in Theory and Practice*. Oxford: Oxford University Press, 2014.

Kelly, Caleb. *Gallery Sound*. New York: Bloomsbury, 2017.

Kim-Cohen, Seth. *Against Ambience and Other Essays*. New York: Bloomsbury, 2016.

Koepnick, Lutz. *On Slowness: Toward an Aesthetic of the Contemporary*. New York: Columbia University Press, 2014.

Koepnick, Lutz. *The Long Take: Art Cinema and the Wondrous*. Minneapolis: University of Minnesota Press, 2017.

Kracauer, Siegfried. *The Mass Ornament: Weimar Essays*. Translated, edited and with an introduction by Thomas Y. Levin. Cambridge, MA: Harvard University Press, 1995.

Kracauer, Siegfried. *Theory of Film: The Redemption of Physical Reality*. Princeton: Princeton University Press, 1997.

Krauss, Rosalind E. *The Originality of the Avant-Garde and Other Modernist Myths*. Cambridge, MA: MIT Press, 1985.

Krukowski, Damon. *Ways of Hearing*. Cambridge, MA: MIT Press, 2019.

LaBelle, Brandon. *Background Noise: Perspectives on Sound Art*. New York: Continuum, 2006.

LaBelle, Brandon. *Acoustic Territories: Sound Culture and Everyday Life*. New York: Continuum, 2010.

LaBelle, Brandon. *Sonic Agency: Sound and Emergent Forms of Resistance*. London: Goldsmiths Press, 2018.

Lacey, Kate. *Listening Publics: The Politics and Experience of Listening in the Media Age*. Cambridge, UK: Polity Press, 2013.

Lawrence, Amy. *Echo and Narcissus: Women's Voice in Classical Hollywood Cinema*. Berkeley: University of California Press, 1991.

Lewis-Williams, David. *The Mind in the Cave: Consciousness and the Origins of Art*. London: Thames and Hudson, 2002.

Licht, Alan. *Sound Art: Beyond Music, Between Categories*. New York: Rizzoli, 2007.

Liu, Alan. *Friending the Past*. Chicago: University of Chicago Press, 2018.

London, Barbara. *Soundings: A Contemporary Score*. New York: The Museum of Modern Art, 2013.

London, Justin. *Hearing in Time: Psychological Aspects of Musical Meter*. Oxford: Oxford University Press, 2004.

López, Francisco. "Profound Listening and Environmental Sound Matter." In *Audio Culture: Readings in Modern Music*. Edited by Christoph Cox and Daniel Warner. New York: Bloomsbury, 2004, 82–7.

Lucier, Alvin. *Chambers: Scores by Alvin Lucier. Interviews with the Composer by Douglas Simon*. Middletown, CT: Wesleyan University Press, 1980.

Lucier, Alvin. *I Am Sitting in a Room*. Lovely Music, Ltd., 1990, CD, album notes.

Lucier, Alvin. *Music 109: Notes on Experimental Music*. Middletown, CT: Wesleyan University Press, 2012.

Macfarlane, Robert. *Underland: A Deep Time Journey*. New York: W. W. Norton & Company, 2019.

Maeder, Marcus, Ed. *Milieux Sonores / Klangliche Mileus: Klang, Raum und Virtualität*. Bielefeld: Transcript, 2010.

Massey, Doreen. *For Space*. London: Sage, 2005.

McCullough, Malcolm. *Ambient Commons: Attention in the Age of Embodied Information*. Cambridge, MA: MIT Press, 2013.

Melissa Gregg and Gregory J. Seigworth, Eds. *The Affect Theory Reader*. Durham: Duke University Press, 2010.

Metzner, Paul. *Crescendo of the Virtuoso: Spectacle, Skill, and Self-Promotion in Paris during the Age of Revolution*. Berkeley: University of California Press, 1998.

Meyer, John M. *Engaging the Everyday: Environmental Social Criticism and the Resonance Dilemma*. Cambridge, MA: MIT Press, 2015.

Misrach, Richard and Guillermo Galindo. *Border Cantos*. New York: Aperture, 2016.

Mockus, Martha. *Sounding Out: Pauline Oliveros and Lesbian Musicality*. New York: Routledge, 2007.

Mondloch, Kate. *Screens: Viewing Media Installation Art*. Minneapolis: University of Minnesota Press, 2010.

Moore, Thomas. *The Re-enchantment of Everyday Life*. New York: HarperCollins, 1996.

Morton, Timothy. "Of Matter and Meter: Environmental Form in Coleridge's 'Effusion 35' and 'The Eolian Harp.'" *Literature Compass*, 5.2 (2008): 310–35.

Morton, Timothy. *Being Ecological*. Cambridge, MA: The MIT Press, 2018.

Mowitt, John. *Sounds: The Ambient Humanities*. Berkeley: University of California Press, 2015.

Munder, Heike. *Ragnar Kjartansson: To Music / An die Musik*. Zurich: Migros Museum für die Kunst, 2015.

Nancy, Jean-Luc. *Listening*. Translated by Charlotte Mandell. New York: Fordham University Press, 2007.

Noonan, Mary. *Echo's Voice: The Theaters of Sarraute, Duras, Cixous and Renaude*. New York: Taylor and Francis, 2014.

Norberg-Schulz, Christian. *Genius Loci: Towards a Phenomenology of Architecture*. New York: Rizzoli, 1980.

Obrist, Hans-Ulrich. *A Brief History of Curating*. Dijon: Les presses du reel, 2008.

O'Doherty, Brian. *Inside the White Cube: The Ideology of the Gallery Space*. Expanded edition. Berkeley: University of California Press, 2000.

Oliveros, Pauline. *Deep Listening: A Composer's Sound Practice*. Lincoln, NE: iUniverse, 2005.

Ovid. *The Metamorphoses*. A new verse translation by Allen Mandelbaum. San Diego: Harcourt Brace & Company, 1993.

Pais, Abraham. *Subtle Is the Lord: The Science and the Life of Albert Einstein*. Oxford: Oxford University Press, 1982.

Pedulla, Gabriele. *In Broad Daylight: Movies and Spectators after Cinema*. London: Verso, 2012.

Pentcheva, Bissera V. *Hagia Sophia: Sound, Space, and Spirit in Byzantium*. University Park, PA: Penn State University Press, 2017

Peters, John Durham. *The Marvelous Clouds: Toward a Philosophy of Elemental Media*. Chicago: University of Chicago Press, 2015.

Pettman, Dominic. *Sonic Intimacy: Voice, Species, Technics (or, How to Listen to the World)*. Stanford: Stanford University Press, 2017.

Price, Peter. *Resonance: Philosophy for Sonic Art*. New York: Atropos Press, 2011.

Rebentisch, Juliane. *Aesthetics of Installation Art*. Berlin: Sternberg Press, 2012.

Reckwitz, Andreas. *Die Gesellschaft der Singularitäten: Zum Strukturwandel der Moderne*. Frankfurt am Main: Suhrkamp, 2017.

Rogers, Holly. *Sounding the Gallery: Video and the Rise of Art Music*. Oxford: Oxford University Press, 2013.

Rosa, Hartmut. *Resonance: A Sociology of Our Relationship to the World*. Translated by James Wagner. Cambridge, UK: Polity Press, 2019.

Rothenberg, David. *Thousand Mile Song: Whale Music in a Sea of Sound*. New York: Basic Books, 2008.

Sauer, Theresa. *Notations 21*. New York: Mark Batty Publisher, 2009.

Schaeffer, Pierre. *Traité des objets musicaux: essai interdisciplines*. Paris: Éditions du Seuil, 1966.

Schulze, Holger. *The Sonic Persona: An Anthropology of Sound*. New York: Bloomsbury, 2018.

Slobin, Mark. "Ensembles—Banding versus Bonding." In *Subcultural Sounds: Micromusics of the West*. Middletown, CT: Wesleyan University Press, 2000, 98–108.

Small, Christopher. *Musicking: The Meanings of Performing and Listening*. Middletown, CT: Wesleyan University Press, 1998.

Snow, Charles Percy. *The Two Cultures*. Cambridge, UK: Cambridge University Press, 1998.

Steege, Benjamin. *Helmholtz and the Modern Listener*. Cambridge, UK: Cambridge University Press, 2012.

Sterne, Jonathan. *The Audible Past: Cultural Origins of Sound Reproduction*. Durham: Duke University Press, 2003.

Stockhausen, Karlheinz. *Texte zur Musik*. Cologne: M. DuMont Schauberg, 1963–.

Strickland, Edward. *Minimalism: Origins*. Bloomington: Indiana University Press, 2000.

Sujahta, C. *Vibration and Acoustics: Measurement and Signal Analysis*. New Delhi: Tata McGraw Hill, 2010.

Szendy, Peter. *Listen: A History of Our Ears*. New York: Fordham University Press, 2008.

Szendy, Peter. *Phantom Limbs: On Musical Bodies*. Translated by Will Bishop. New York: Fordham University Press, 2016.

Taussig, Michael. *Mimesis and Alterity: A Particular History of the Senses*. New York: Routledge, 1993.

Terkessidis, Mark. *Kollaboration*. Frankfurt am Main: Suhrkamp, 2015.

Toop, David. *Ocean of Sound: Aether Talk, Ambient Sound and Imaginary Worlds*. London: Serpent's Tail, 1995.

Toop, David. *Sinister Resonance: The Mediumship of the Listener*. New York: Continuum, 2011.

Trower, Shelley. *Vibration: A History of the Pleasure and Pain of Sound*. New York: Continuum, 2012.

Tuhus-Dubrow, Rebecca. *Personal Stereo*. New York: Bloomsbury, 2017.

Uroskie, Andrew V. *Between the Black Box and the White Cube: Expanded Cinema and Postwar Art*. Chicago: University of Chicago Press, 2014.

van Alphen, Ernst and Tomáš Jirsa, Eds. *How to Do Things with Affects: Affective Triggers in Aesthetic Forms and Cultural Practices*. Leiden: Brill Rodopi, 2019.

Voegelin, Salomé. *Listening to Noise and Silence: Towards a Philosophy of Sound Art*. New York: Continuum, 2010.

Weber, Max. "Science as Vocation." In *From Max Weber: Essays in Sociology*. Edited by H. H. Gerth and C. Wright Mills. Oxford: Oxford University Press, 1946, 129–58.

Weizman, Eyal. *Hollow Land: Israel's Architecture of Occupation*. London: Verso, 2007.

Wiens, Kathleen and Eric de Visscher, Eds. *Curator: The Museum Journal*, 62.3 (July 2019). Special Issue on the "Sonic."

Zeitlinger, Anton. *Dance of the Photons: From Einstein to Quantum Teleportation*. New York: Farrar, Straus and Giroux, 2010.

Zisiou, M. "Towards a Theory of Museological Soundscape Design: Museology as a Listening Path." *Soundscape*, 11.1 (2011): 36–8.

Index